INTERNATIONAL ENERGY OPTIONS

The full programme of the 1980 Conference is given as an Annex. The 54 papers listed are held in the archive of the British Institute of Energy Economics.

Paul Tempest is the Bank of England's energy specialist and arabist. Since 1963, he has also worked on secondment with the Bank for International Settlements, the Middle East Centre for Arabic Studies, the Qatar and Dubai Currency Board and Shell International. He is currently chairman of the BIEE and one of the two UK Council members of the IAEE.

International Energy Options:

An Agenda for the 1980s

Selected Papers from Leading
Contributors to the 1980 Annual
Conference of the International
Association of Energy Economists in
Cambridge, UK

Edited by
Paul Tempest

Oelgeschlager, Gunn & Hain, Publishers, Inc.
Graham & Trotman Ltd, Publishers

Published jointly in 1981 by:

Oelgeschlager, Gunn & Hain, Publishers, Inc.   Graham & Trotman Ltd
1278 Massachusetts Ave   Bond Street House
Harvard Square   14 Clifford Street
Cambridge   Mayfair
MA 02138   London W1X 1RD
USA   UK

Available in the USA and Canada from Oelgeschlager, Gunn & Hain, Inc. and in
the Rest of the World from Graham & Trotman Ltd.

Library of Congress Cataloging in Publication Data

Main entry under title:
International Energy Options: An Agenda for the 1980's
Includes index.
1. Energy policy — Congresses. 2. Power resources — Congresses. I. Tempest,
Paul. II. International Association of Energy Economists.
HD9502.A2T68   333.79   81-80575   AACR2
ISBN 0-89946-089-5 (Oelgeschlager, Gunn & Hain)

©International Association of Energy Economists, 1981

ISBN: 0 89946 089 5 (Oelgeschlager, Gunn & Hain)
      0 86010 303 X (Graham & Trotman)

Typeset in Great Britain; printed in the Federal Republic of Germany.

Contents

v

Introduction

Paul Tempest

"The economic issues that have dominated our thoughts are the price
and supply of energy and the implications for inflation and the level of
economic activity in our countries and for the world as a whole. Unless
we can deal with the problems of energy, we cannot cope with the other
problems........ "

Communique of the 1980 Venice Summit Meeting (opening words)

This collection of distinguished contributions to the current energy
debate has been drawn from the proceedings of the 1980 annual
meeting of the International Association of Energy Economists which
met, concurrently with the Venice Summit Meeting, in Cambridge,
UK, in June 1980.

The 1980 Conference was attended by 248 energy economists
from 16 countries, representing a wide spread of government, industry,
finance and academic viewpoints. Among the 54 invited speakers, a
balance was sought between producer and consumer interests, between the industrialised and developing countries and between
longer-term research and more immediate issues. This collection
makes no such claim to cover the field in any comprehensive manner.
All the detailed technical work of the Conference has been omitted.

Rather, we have tried to pin-point, in the selection of abstracts, those areas where the conventional wisdom concerning energy supply and demand is beginning to change, where some major break-throughs in energy supply and energy efficiency can be expected, and where centres of excellence can be identified for support and development to maximise their impact. The juxtaposition of these papers, which represent some of the most authoritative and up-to-date statements on their particular areas of concern, tends to over-emphasise differences of opinion. None the less, only in the final sections is an attempt made to assess the extent of consensus in the main areas of debate.

The papers have been arranged to provide a thread from a basic diagnosis of current energy problems, through an analysis of govern-mental and commercial response, to the constraints and opportunities for capital and labour. Contingency planning for supply shortages and prospects for increased efficiency are then examined in some detail and lead to the conclusion that our current energy problems are quite soluble, even within our current technological competence. In the longer-term, the breakthrough to stable energy supply and price, as a prerequisite for investment and economic growth, is already clearly within our reach, provided adequate action on an international scale can be mobilised and harnessed in time.

THE GLOBAL DILEMMA

Despite an underlying abundance of energy, immediate short-falls of energy supply present a fundamental dilemma. The very low expectations of economic growth induced by these shortfalls together with, suddenly, excess capacity in the production of energy, caused by an abrupt fall in demand, offer major deterrents to investment. The problem is therefore essentially one of investor confidence. If, despite current gloom, confidence can be restored, it will be self-fulfilling. If not, national and commercial competition will waste effort in chasing ever-declining supply. To what extent can we rely on the invisible hand of market forces to supply a solution, when the market is itself fragmented and partly dominated by monopoly interests? How indeed can a steady rise in the real price of energy provide incentives to alternative investment when the industrialised world is crippled by the progressive transfer of resources to OPEC, and when the rest of the world is becoming too impoverished to con-tribute? As our two contributors to this section, Sir David Steel and Chauncey Starr point out from widely different viewpoints, what is at risk here is that power politics and national opportunism could conspire to frustrate the working out of economic and market forces.

CAN GOVERNMENTS RESPOND IN TIME?

We are very pleased to be able to include in this section the statements by John Sawhill, then Deputy Secretary at the United States Department of Energy (and subsequently appointed by President Carter as Head of the US syncrude project), Norman Lamont, Parliamentary Under-Secretary at the UK Department of Energy and Dr Ali Attiga, Secretary-General of the Organisation of Arab Petroleum Exporting Countries. All three contributors stress the risk of permanently damaging the global economy and all three conclude that the problem is essentially international in character and must ultimately be resolved by international consensus and co-ordinated action. Government response at a purely national level, which ignores the wider dimension of the problem, may in fact constitute one of the greatest risks.

ENERGY IN WORLD POWER POLITICS

Michael Kaser's authoritative review of the East-West energy balance provides considerable reassurance that—at least in the 1980s—any net emergent demand for Western energy by the USSR and satellites is unlikely to be on a very large scale, although the energy supply problems of East (and West) Europe will be quite complex. The danger, as Melvin Conant has pointed out here and elsewhere, is not so much outright super-power confrontation on energy supply, but rather the dangers of divisions and consequent weakening within the Western camp, if equitable and positive solutions to energy problems cannot be found.

CAPITAL CONSTRAINTS AND OPPORTUNITIES

Two areas in which the West's performance has consistently excelled, when compared with that of the USSR and satellites, are first, energy technology and second, the flexible mobilisation of industrial finance. As far as the industrialised world is concerned, there is no overall shortage of capital for investment in new energy. The experience of financing high-cost energy such as North Sea oil and gas and nuclear electricity have established proven methods for Governments to underpin commercial risk through guarantees, assurances and participation. It is very difficult therefore to conclude that lack of capital will constitute a constraint on energy development; indeed, such large-scale development in the industrialised world is much more likely to provide large-scale financial opportunities, and conceivably, a motor for a resurgence of industrial investment in general.

The financing problems of the less-developed world are much more complex. From the twelve papers and two sessions devoted to the developing world, we have selected Dr. Pachauri's statement, which points out that, whereas the non-oil developing world coped relatively successfully with the 1973 oil price rise mainly by increased capital market borrowing, other mechanisms will have to be found, as more and more developing countries begin to press against their commercial ceilings on the international capital market. The need for specialised agencies and co-ordinated international action are particularly emphasised.

NEW APPROACHES TO PRODUCTIVITY AND EMPLOYMENT

As Professor Jorgenson demonstrates in a major contribution to this volume, the economic slowdown in 1974/6 in the United States had its principal origin in declining productivity. The increased relative cost of energy in all the industrial sectors examined proved the one factor, for which there was no ready reaction, substitution or compensation. The immediate prospect of steadily rising domestic energy prices in the United States poses serious doubts for productivity growth and therefore for economic growth in general. The price of capital and labour are, however, susceptible to Government action. Professor Jorgenson concludes therefore that, in the absence of Government measures to cut taxes on capital and labour inputs, the performance of the US economy during the 1980s could be worse than during the period 1973–1980. Increasingly, as increased energy costs to industry feed through the industrialised economies of the West, the pressure to re-examine selective tax alleviation is strengthened. There is no doubt that the relative weight of taxation on different forms of energy has changed markedly over the years; the relative weights of the various fuels have also changed, for mainly different reasons. It is, therefore, highly unlikely that the distribution of domestic taxation of energy bears much relationship to the strategic and economic priorities of each government in determining its optimum energy mix.

The energy-output coefficient, the ratio of energy consumption to the level of output, is generally taken as a prime indicator of the energy price mechanism and as a measure of energy conservation. Professor Ernest Berndt, who has been prominent in this area of energy economics for some time, warns in a joint paper with Dr Watkins of Datametrics against the dangers of misunderstanding raw energy-output coefficients. Taking the Canadian textile industry in 1975-76, they demonstrate the complexities behind the simple ratios and the need for both a sophisticated dynamic model and adequate data to provide a convincing economic rationale. In the case of Canadian textiles, they conclude, beyond doubt, that contrary to the apparent evidence from superficial energy-output coefficients, rising energy prices had in fact effectively restrained energy consumption.

The third major contribution selected for this section is the case argued by Arnold Packer, Assistant Secretary in the United States Department of Labor, and Wilbur Steger for a possible route out of current economic stagnation and damaging levels of inflation. The vicious circle of high inflation induced by energy price jumps followed by harsh, deflationary policies of consumer governments, can only be halted by changes in expectations. Nothing, it is argued, would be as effective in this regard as a massive government commitment to an international high-technology investment programme: a necessary corollary is the weaning of government from assistance to traditional and declining industries and a monitoring of the employment and environmental consequences of this process.

NEW INSIGHTS INTO ENERGY DEMAND

Increasing awareness of insecurity in imported energy supply has only recently reversed the consumption trends of the previous two decades. While road, sea and air transport appear bound to oil for the foreseeable future, rail transport and most other uses of oil offer scope for fairly rapid substitution, even at present oil-price levels. Aart Beijdorff of Shell International concludes in this volume that the energy market will react very quickly to price signals and that given current industrial depreciation and renewal practice, very fundamental savings will be achieved over the next 15-20 years. Professor Patrick O'Sullivan of the Welsh School of Architecture and

Frederic Romig reinforce this opinion, pointing out the potential for interfuel substitution in the industrialised countries and the gradual emergence of much more flexible consumption patterns.

ASPECTS OF NEW ENERGY SUPPLY

Out of numerous studies of new and potentially significant sources of energy, we have selected authoritative statements on the potential of the Orinoco tar sands, an assessment of progress in the French nuclear programme and a study of bio-mass in the United States.

OPPORTUNITIES IN THE
TRANSITION PHASE

The urgency of transition to new forms of energy turns largely on the speed with which reserves of oil and gas are depleted. Given the steady increases in oil recovery rates (still, overall, only about 30%) and progress in gas utilisation and the exploitation of smaller and smaller oil and gas deposits, we have every chance of spinning out the oil and gas resource well through the next century. Although often challenged by the major oil companies on his assumptions, Professor Peter Odell of Erasmus University has argued consistently over the last decade that oil and gas reserves are larger than acknowledged by the industry or governments. Upward revisions of official estimates have until recently tended to prove him more right than his critics. In the paper in this volume he sums up his principal argument.

RISKS AND THE NEED FOR
CONTINGENCY PLANNING

Given the apparently wide consensus that the risk of interruption to oil supply is likely to be high for the foreseeable future, and the equally wide acceptance of the damage caused by supply and price discontinuities, much official and international effort has been poured into contingency planning. Of the three contributions selected for this section, Michael Telson argues strongly against US Government policies in 1979-80 which tended to exacerbate local shortage and encourage hoarding, Fritz Lucke registers an equally strong defence of market forces in Germany, whereas Oystein Noreng takes an intermediate position.

TOWARDS THE INTERNATIONAL
ENERGY BREAKTHROUGH

It is becoming clear that the breakthrough to energy supply which matches our needs and is available and in prospect at a stable real price will come most probably as much on the demand side as on the supply side. Both sides now offer considerable hope from new technology. Finance does not appear to present an obstacle. In this final section, Christopher Johnson of Lloyds Bank stresses the need to sustain global economic growth and Len Brookes outlines the essential role of nuclear power. Finally, in the closing address of the Conference, the current President of the International Association of Energy Economists, Professor Adelman of the Massachusetts Institute of Technology, sums up with an agenda for policy decision and research in the 1980s.

Paul Tempest,
Chairman, British Institute
of Energy Economics,
9 St. James's Square,
London SW1Y 4LE

PART I

The Global Dilemma

Chapter 1

Risks in the International Oil Trade

Sir David Steel*

The theme which I would like to discuss is the relationship between security of supply and international trade in energy. There are costs, benefits, and risks in both security and trade. The values of all three are changing fast. New balances are being struck, especially by governments, whose intervention in these matters becomes every day more obvious and less simple. I believe economists can make a great contribution to the discussion of this subject, because the evaluation of costs, benefit and risk is your professional concern. As "energy economists" you also know how important are the physical constraints which the hard facts of nature put upon us. What can be done, how quickly, and where, are not trivial problems in technology, nor should they be in the economics of energy.

The most important and hardest fact in the energy scene is that from a tangle of geological and economic history we have today a world in which the geography of energy consumption is very different from the geography of energy production. The difference is made up by trade. Traded energy—most of it oil—supplies 30 percent of the consumption of the OECD world and disposes of 73 percent of the production of the OPEC members.

*Chairman, British Petroleum

Oil has been a case study in the classic benefits of international trade. There were overwhelming comparative advantages of producing oil in the Middle East, and of manufacturing goods in Japan, Europe, and the United States to exchange for that oil. The international oil companies helped to create this trade and profited from it. We could feel our profits well earned because the benefits of interdependence were so great.

I shall not discuss in detail what brought the expansion of that trade to a halt. You all know how the rapid depletion of the world's known reserves precipitated a crisis of confidence about future oil supplies; how political changes in the Middle East in particular precipitated the change in the international companies' role there; and how unexpectedly sudden and steep recent changes in the price of oil have been.

Right now, the international trade in oil appears to confer economic benefits on exporters which are limited only by their capacity to absorb them. The benefits to importers have been cut back, crisis by crisis. Heavy fuel oil prices have now passed the probable cost of developing alternatives such as coal and nuclear heat for electricity generation and heavy industrial use. Distillate product prices may be within whispering distance of some estimates of the cost of synthetic liquid fuels—at least in places where coal or other feedstocks are in surplus.

Demand for oil is falling. Between 1973 and 1979 in the European Community GDP grew by 15 percent overall, consumption of oil fell by 6.7 percent. Now we have had another year of price increases. Worldwide, it is possible that industrial countries' oil consumption has now at least levelled, and possibly has begun a falling trend. There may be some increased demand from developing countries, and perhaps from the Soviet bloc. Oil production outside OPEC may, with great efforts by all concerned, be held up near its present level to the end of the century. While the oil trade may level off, trade in gas, coal and uranium is likely to expand—possibly very fast.

I hesitate to go from these ill-defined trends to any kind of specific forecast. Some of you here may be bolder than I. My point is simply that although the international oil trade has probably stopped growing, it is not going to disappear or even decline very much for some time to come—a couple of lead times at least.

We have therefore to consider the conditions under which this trade will take place. I will not discuss the structural changes of the last year, which I think other speakers will deal with.

The point I would prefer to dwell on is that this continuing international oil trade, with all its economic benefits, has obviously become very much more risky than it has ever been except in time of war.

Traders, whether large or small, are at risk because sudden changes in price are now common, and the value of even a small cargo is very high. The risks are high, but so are the rewards. Economists will not be surprised that the number of traders has increased dramatically.

Importing refiners are also at risk because of price fluctuations. Their risks are compounded because of fluctuations in availability of particular grades. As contract terms have become shorter, forward planning, whether of programmes or of plant, has become riskier. Since refiners' margins in many countries are restricted by price controls, I am afraid that the number of refiners, or at least the scale of their operations, will diminish. Not many willing risk-takers there!

Distributors and retailers get some additional risks for the same reasons, with the added complication that the effect of price upon their customers is so new as not to be well known. Product dealers may have to learn the hard way how fast their market is shrinking— unless perhaps some economists can help them out.

Consumers are also troubled by price, whether oil or gas price. But for them perhaps the greatest risk of all is discontinuity of supply. Vehicles without gasoline will not move. Factories without fuel will stop producing. People without heat in winter will die. Because of the seriousness of these risks governments of consuming countries are bound to be politically concerned about the security of supply.

Uncertainty is just as inevitable at the other end of the business— in countries whose economies are wholly or mainly dependent on the export of petroleum. This was the original qualification for OPEC membership. Though the terms of trade have shifted in favour of these countries it is not all smooth going for them. Though the changes are working to the general advantage of oil exporters as a bloc, the position of an individual exporting country is not so certain: is it depleting too fast in relation to its reserves? Is it spending its revenues wisely? Are its foreign assets secure? Is their real yield certain? Some of us may feel that those risks are being reasonably well rewarded.

There are more serious questions. Many of the oil exporters are countries with small populations, few other resources, and limited technological capacity. Their independence and progress depends on the political balance and prosperity of the world outside their borders. How far is that balance being upset by successive oil crises?

So it seems that, although international trade in oil is bound to continue, it has become more risky, or creates more risks, for every-one. How will the system adjust to this?

The first and most obvious response is that the value of indigenous supplies of energy is enhanced because they are free of this particular set of risks. This does not automatically make them all competitive: they may be expensive, like shale oil. They may be difficult to organise because of internal political conflicts about the benefits of ownership—like North American oil and gas; they may involve environmental risk, like coal; they may involve other risks which are difficult to calculate, like nuclear power.

New domestic oil and gas production is still encouraged in most importing countries, most of the time. The countries with the best geological prospects, as most of you know, also elect and employ the most dedicated tax gatherers. The wedge that is being driven between the price the consumer pays and the price the producer gets is now getting to be of an uncomfortable size and shape. Such taxes keep producer prices well below consumer prices. If the taxes increase more than proportionally whenever the consumer price rises, the signals of scarcity are suppressed. Moreover, even when prices are steady, costs are not. In some countries the price the oil producer is paid after tax, royalty, and government participation does not adequately cover the risks the producer is required to take as costs rise. It is quite possible therefore that the development of new oil production capacity outside OPEC over the next few years will fall short of what is technically possible, of what the Governments concerned expect, and what is justified by the price consumers are already paying.

Whatever the incentive, few major industrial countries can seriously hope for "energy independence" within one or even two lead times of real projects. Britain is probably the exception. Most will remain net importers of oil, albeit in decreasing quantities. For many countries such as Japan, and most European countries, gas, coal and uranium imports will go up as oil imports level or fall. Though the United States may progress towards "energy independence" it is likely to remain an importer of oil for a good many years. We hope the United States will also become a growing exporter of coal to the international market. These new international energy trades have their risks too. The timing of supply facilities with long lead times is unlikely exactly to match the speed at which users switch from oil to the alternatives. Mismatches will cause short-term feast or famine, and the timing of individual projects is therefore commercially risky.

The second response to the increasing risk of international oil trade is that risks will be shifted through the market to firms and individuals who are prepared to take them. Commercial companies like my own typically take almost every kind of risk: trading risks, refiners' risks, and dealers' risks. We also take the exploration and development risk—and the risk of further taxation—inevitable in major oil field development projects. One of the greatest worries is that Governments will change their policies, or their instruments of policy, in unexpected ways. As economists, you will not be surprised if I dwell on the fact that we take risks in the expectation of reward. Our readiness to do so becomes eroded if those expectations are repeatedly disappointed because fiscal action, price control, or other governmental intervention comes between the risk we take and the reward we expected.

But others are coming into the game—notably into oil trading. Trading channels are becoming more diversified and it may be that patterns will evolve which are more stable than we are now experiencing. It is right to think constructively about adopting institutions and practices—such as the spot market. The recent proposal by the London Commodity Exchange for a properly organised futures market in Europe in some oil products certainly deserves study and my company is co-operating with that study.

A third approach to the new risks in the international oil trade is to involve importing Governments directly in the oil trading business. This is probably inevitable where, as in the OPEC countries, the economy is mainly or totally dependent on the international oil trade. The case is not so obvious elsewhere. Bilateral trading between Governments notoriously introduces rigidity and inefficiency into the distribution system. These impose real costs. How can Governments of importing countries secure cheaper or more secure oil supplies from the international market than the variety of commercial channels can provide? There seems to me to be three possible, but not very convincing, answers.

Firstly some Governments, notably of some developing countries, appear to have obtained cheaper oil because it suits the political objectives of the oil exporting country to provide it. This creates a new kind of political dependence which cannot be without cost to the importer and is certainly not without risk.

Secondly, some governments—including some governments of industrialised countries—appear to have obtained extra oil in times of crisis by paying more for it. Governments have a facility for recovering their costs which is not open to commercial firms. When Governments "scramble for oil"—as some did in 1979—the bidding is likely to go higher as a result. Their intervention raises the general price level and therefore imposes costs on importers everywhere.

Thirdly, some governments can organise commercial packages which link oil imports and industrial exports. If that is the way in which some of the exporters are determined to do it, that is the way their trade will tend to be. It could be better to have inefficient, inflexible and costly international trade than no international trade at all. Not many of the oil exporting countries are eager to forego the advantages of exchanging oil for that equally fluid commodity—money.

A final response to the increasing riskiness of international oil trade might be to seek some form of international agreement between importers and exporters. Various ideas have been canvassed. Detailed mechanisms have been proposed which have considerable rational appeal—and I expect there are some rational people here today. There are fundamental difficulties about most such schemes which perhaps explain why they are not instantly adopted. Such international agreements depend on the mutual balance of interests being demonstrated all the time. There is no enforcement, Governments tend not to bind their successors and Governments can change unexpectedly. This immediately restricts the practical scope of effective oil trading international agreements to those where the benefits on both sides are delivered in a fairly short term, under all the changing circumstances of day to day.

To sum up: the changes which have taken place in the oil market greatly increase its riskiness. International trade in gas, steam coal, and uranium is likely to increase, but there is great uncertainty about precisely how and when. Though the risks of the international energy trade, especially oil, have become very great, the benefits are also still very large, even for imports. So the oil trade will continue and other energy trade will grow. To reduce the risks that now accompany international trade most countries will try to boost indigenous supplies even at some extra cost. Such efforts will fail if producer prices after tax do not adequately cover rising costs and the technical and financial risks which explorers and developers must take. Some of the new risks will eventually be spread through the market, maybe including some new mechanisms. The scope for reducing the cost or increasing the security of international oil supplies by individual Government trading is limited: the stability of Government-to-Government oil deals has yet to be proved—linking oil trade with other matters may be either politically expensive or commercially risky, or both.

I have talked mainly about international oil trading. It takes place in a wider energy economy. The riskiness of energy trade is important because it is still so large. There is so much uncertainty about the rate at which alternative energy supplies will develop and growth in demand for energy will fall in response to price. If we cannot easily see how to reduce the *riskiness* of international oil trade we can at least see how its *importance* can be reduced. Clear and efficient policies for the pricing, development, and use of alternatives will not affect the supply of energy for some years, but they will affect the expectations of oil producers today: their depletion and pricing policies are not entirely opportunist. Their attitude to trading the oil in the ground cannot fail to be affected by their idea of its future value. This will be influenced by the commitment of the importers to the development of alternative supply.

Even in the short run, some of the demand pressures on the oil market would be reduced if more Governments were more willing to let consumer prices respond quickly to the signals of scarcity.

I know that it is easier to prescribe such policies than it is for Governments to carry them out. What I have described is a formidable agenda for action.

I doubt if any one has all the answers, but everyone has some stake in the result—even if the stake is purely an intellectual one. Events are moving with great speed. The time for deliberation is short.

Chapter 2

Energy at the Crossroads: Abundance or Shortage

Chauncey Starr *

In words written 150 years ago, Thomas Macaulay, the eminent Cambridge historian provided an apt commentary on today's prevailing mood toward energy:

> "We cannot absolutely prove that those are in error who tell us that society has reached a turning-point, that we have seen our best days. But so said all who came before us, and with just as much apparent reason.... On what principle is it that, when we see nothing but improvement behind us, we are expected to see nothing but deterioration before us?"

The attitude which sees "nothing but deterioration before us" is echoed endlessly in much that is now written about the outlook for energy in the United States and throughout the industrialized world. If this were merely a perception, it would not be a source of much concern, for, in that case, the future would take its own course and today's dire predictions would eventually be forgotten. The real danger is that in large measure our energy future will be the result of the policies we pursue, and policies, in turn, are

*Electric Power Research Institute, Palo Alto, California

the products of prevailing beliefs about the potentials that we foresee.

Are we, in fact, facing a fundamental structural change in the conditions of energy supply with which the world will be forced to live for evermore? I do not think so. I do not believe that the world is imminently running out of its fuel supplies, not even its supplies of mineral fuels, nor do I think that we are approaching pollution limits imposed by the carrying capacity of the natural environment.

We are, however, in a period of energy transition—a transition away from the transient bonanza which has been provided by the world's low-cost petroleum and natural gas, on which all of us have become increasingly dependent since the end of World War II. The new circumstances toward which we now must move, and whether they are to be characterized by abundance or shortage, is up to us. These fateful decisions are now truly in our own hands.

The past economic growth of the industrial nations has been nurtured by a readily available and low-cost energy supply which stimulated a high production efficiency based on energy-intensive machines, a rising domestic standard of living, and intensive market competition which rewarded the efficient use of the world's resources. Because of the worldwide increase in basic energy costs during the past decade, the industrial nations are now faced with two alternatives: either to accommodate to a perceived shortage and plan an energy-limited society; or promote the intensive exploitation of new energy sources to assure availability, even though such new sources may be more costly than today's supply.

The usual initial response to increased energy costs is to seek a reduction in demand. Economic forces will generally cause some shift to more labor-intensive production. But an excessive substitution of labor for advanced energy-using production machinery, while it may provide employment, does so at the price of wasting society's productive resources. This would result in a reduced average standard of living, and is a poor way to save energy.

Conservation is an obvious demand-reducing response to high energy costs, but its mode of implementation depends on whether we face a shortage or abundant supply at higher unit costs.

Survival in an energy-deficient world generates political pressures for intervention which will undoubtedly lead to an administratively controlled society, with energy price ceilings and mandatory limits on energy-intensive activities, such as already are being promulgated in the United States, as, for example, requirements on automobile fuel performance, speed limits, room temperature limits, scheduled days for buying gasoline, rationing of fuel and leisure activities on the horizon, and restrictions on home and building design. This is the leading edge of the "centrally planned lifestyle"—the regimented police state so abhorred by those of a free spirit.

In contrast, if a successful effort is made to promote energy abundance, albeit at higher cost, conservation would follow naturally as a result of higher prices. Each user will choose to adjust his consumption pattern and selection of energy-efficient systems to match his own resources and needs, but the user will not face unavailability. Abundance would result, for example, if we stimulated technologies on hand, such as shale oil recovery, coal conversion to liquid fuels and gases, nuclear power, and the numerous competitive low-temperature solar alternatives. Every analysis indicates a plentiful mix from these sources—certainly sufficient to guide our energy transition steps and to avoid serious limitations in the foreseeable future.

Thus, we are at a key crossroads on energy policy: either we plan to accommodate to enduring shortage; or, alternatively, to vigorously promote the development of available energy sources. Higher-cost abundant energy will not have the same result as a truly constraining energy deficiency. Living within an affordable energy supply is different than hand-to-mouth survival on short rations.

The issue goes to the root of our societal perception of the future. Either we foresee a future of shortages arising from the "limits to growth" in a finite earth; or we foresee an expansionist future made possible by the "growth of limits" resulting from the advancement of technology's frontiers. Historically, the world has always been faced with pending resource limitations, and technology has been able to expand these limits to meet society's needs. This faith in technology's progress is more than wishful dreaming—many of us see the paths ahead that may lead us to such outcomes in the energy areas.

Let us take the energy situation in the United States as a case study of these issues. In using the United States as an example, I am, of course, dealing with the situation I know best. But there is more to it than that, because what happens in the United States produces worldwide repercussions. Just as the precipitate movement of the United States away from essential energy self-sufficiency, mainly during the past decade, has been a major factor in producing today's worldwide fuel disorder, so a deliberate movement of the United States towards greater self-sufficiency in the future could have profound effects in introducing a new stability into world energy markets.

A key issue for national decision makers is the anticipated growth in future energy consumption. We take a 20-year horizon for illustrative purposes. In the interconnected net of social expectations, economic growth, and energy demand, let us initially enter with a projection of minimum social expectation, namely, that each worker in the 1980-2000 labor force will expect the same real income over his lifetime that a person of equivalent education and experience would receive today. Thus, as a minimum, each worker would look forward to the same time pattern of real income for himself as he sees today in other workers of comparable ages, skills, and education. Empirical values for this concept of minimum expectations have been derived by the EPRI staff. The forecasted increase in the work force from 1980-2000 is about 35 per cent, most of whom are already born. Because of the age cohort distribution, the average age of the work force in 2000 will be six years older. At that time the educational level of the work force will also be higher: college-educated, 25 percent then versus 16 percent now; high school-educated, 60 percent then versus 50 percent now. These changes raise the average individual real income by about 25 percent, and the aggregate personal income about 66 percent, or 2.6 percent/year growth rate. If the ratio of personal income to GNP remains roughly constant, the U.S. economy needs to grow at about the same rate to meet these likely labor force minimum expectations. Increased transfer payments to the larger fraction of retired workers will certainly push GNP goals above this 2.6 percent/year.

A historical check shows that the GNP growth rates have always managed to provide for more than minimum expectations, except during the Great Depression period of the early 1930's and the post-1973 embargo recession. It is doubtful whether the present U.S. social fabric can survive with such a "zero expectation growth" in the coming decades. The lowest income sector of our population has been demanding a major improvement in its status, so our minimum social goal is likely to exceed this 2.6 percent/year. How much is, of course, a political uncertainty, as is the maximum economic growth rate that we can pragmatically expect to achieve. For reference, the 1950-1960 GNP growth rate was 3.2 percent/year, and the 1962-1972 rate was 3.9 percent/year.

What does this GNP projection tell us about total energy demand? There is not the time here to explore the intricacies of energy-GNP relationships, so I will not develop the detailed analysis—it is available in EPRI staff studies. Our findings are not startling. Taking into account a range of economic growth of 2.5-3.5 percent/year, and an expectation of substantial conservation, the projected range of total annual energy demand in 2000 is from about 103 quads to 138 quads, with a present base of about 80 quads, or roughly from 1.3 to 1.8 times our present level.

Electricity growth projections are even more uncertain, because they involve the additional factor of interfuel substitution between electricity and other energy forms. Electricity has always grown at a faster rate than total energy, and the EPRI study projects electricity growing about 1.5 times as fast. For the year 2000, this results in electricity production about 2 to 2½ times the annual production today.

Where will our energy supplies come from to meet these needs? Our national energy problem has usually been presented as one of oil availability, import price, and our resulting negative balance of trade. The understandable public interest in assuring a continuing supply of liquid fuel for private autos and transportation services has resulted in focusing political attention and national planning predominantly on this issue. This topic has been so often reviewed that I will spare you the tedium of listening again to the litany of oil trade numbers and their financial and political implications.

But I do want to make my point that technology can substantially alter the liquid fuel future. First, accelerated near-term electricity production from solid fuel—coal and nuclear—could eventually displace a portion of oil use in three areas: under power plant boilers (9 percent of total oil use); for space and water heating (12 percent); and industrial use (21 percent). The target here (42 percent) is almost as large as all U.S. oil imports. Additionally, electricity can often substitute for gas, which, in turn, may substitute for oil as a feed stock for the petrochemical industry.

How many electric power stations would all this take? If heat pumps are used to efficiently convert electricity back to heat, a rough substitution equivalent is 30,000 barrels of oil/day for each nominal 1,000 MWe power station (coal or nuclear). The United States now uses about 18 million barrels/day, so 250 power stations could, in principle, replace all the above 42 percent. We now have operating the equivalent of about 580 such stations, so an increase of less than half of our present capacity would fulfill this need, assuming such optimistically complete and efficient substitution would actually take place.

Second, oil recovery from oil shale is potentially extremely large, and could replace all our imported oil in time. This is a production process that is inherently more costly than naturally flowing wells, but the costs are within near-term commercial reach. What is needed now is the large-scale demonstration that settles all the ancillary uncertainties—environmental issues, equipment reliability, fuel refining variables, etc. Oil shale has a near-term potential that should be obvious to everyone.

Third, coal conversion to liquids and gases is a chemical engineering process which has been feasible on a small scale for some time, but large-scale engineering development remains to be accomplished. This might be done during the coming decade, if proposed projects are started soon. Cost projections are promising, but very uncertain, because the large-scale engineering subsystems have not been demonstrated. The very large coal resources of the United States make such conversion potentials a very attractive long-term objective for liquid fuel supply.

Thus, there is no question in my mind that technology is showing us a way to remove the oil squeeze, with room to spare. We can be optimistic about an eventual assured liquid fuel supply, if we take the obvious steps.

Crucial as oil supply may be, there is another portion of the U.S. total energy mix which may be less obvious, but equally important—the supply of electricity. At present, almost one-third of our primary fuel equivalent is used for the generation of electricity, and by the year 2000, about one-half of our equivalent primary energy input will be so used. Of the electricity we generate, only a third is for residential use—two-thirds is used by the business sector (commercial and industrial). I would like to focus on this portion of our future energy supply problem in the short time available here.

The electric utilities are generally in a good position to supply the U.S. needs in the near term. New generating capacity, committed in the halcyon days of economic growth of the 1960s and early 1970s, has been steadily coming on-line. With the reduced economic growth since 1973, this new capacity provides a reserve that can take care of the increasing needs of the next few years, with some pending regional shortfalls. But what about the long term? With a ten to twelve-year time lag between decision and availability, deficiencies in electricity supply are not easily rectified.

A prime threat to the secure supply of electricity in the future is the *de facto* moratorium that exists on the growth of nuclear power. Nuclear plants have represented about half of all planned capacity in the United States. Although the nuclear power issues have generally been considered as problems for the electric utility industry, their implications are crucial to the nation's economic future. What is at issue is the future availability of sufficient energy supply to permit the production of industrial goods and services under realistic projections of the growth of the national economy.

How can a doubling of electricity output be realized by 2000? Obviously, the bulk must come from coal and nuclear. Coal power plants now supply about 41 percent of our electricity needs, and consume 480 million tons of coal annually in the process, two-thirds of the total production of coal. Considering all the constraints of environmental regulations on end use, and the delays and institutional constraints on increasing coal supplies, by 2000 coal-produced electricity may be realistically limited to slightly more than double that of today (a 5 percent/year growth), perhaps providing a 45 percent share.

The increasing difficulty of building coal power plants is not generally understood. From time of decision to availability now takes eight to ten years, of which about half is used for the approval chain. For new coal stations designed to meet the projected environmental requirements, the estimated cost of environmental controls is about 60 percent of the total, and, as a result, the total cost of a coal station now almost approaches that of a nuclear. Coal fuel costs are, of course, much higher than nuclear. Coal expansion may be further constrained, if it is extensively used to make synthetic liquid fuel. One ton of coal may hopefully convert to three barrels of oil. So half of our present coal production converted to oil would produce about one billion barrels of oil per year—about 16 percent of our present use and about a third of our imports.

Hydroelectricity provides about 11 percent of our generation output now, and could possibly be increased during the next twenty years, but not to the extent of doubling, so that it may provide about 7 percent of the output in 2000. Geothermal supplies are about 0.2 percent now, and hopefully will be about 2 percent in 2000.

With regard to the solar contribution, the February 1979 report to the President of the *Domestic Policy Review of Solar Energy* presents an interagency forecast of the year 2000 contribution of solar electricity (thermal, photovoltaic, and wind) based on extremely optimistic (in some cases unrealistic) assumptions of successful technical development. Their projection for an equivalent energy displacement by solar is 2-6 percent of the minimum 2000 electricity demand we project.

The total from these estimated sources is 56-60 percent of that required to meet the low projected need for the year 2000. The remainder must come from nuclear, synthetics, oil, and gas. At present, about one-third of our electricity generation depends on oil and gas. Given our national need for liquid fuels for transportation and the strong federal policy to diminish their use for electricity generation, it is unlikely that synthetics and new oil and gas can be assumed to be available for electricity growth. Oil and gas will probably generate somewhat less electricity than today, perhaps 13 percent of the year 2000's minimal needs. Thus, we are left with about one fourth of our needs unfilled, even with our minimum growth estimate, and nuclear is the only source that can fill this gap. Recognizing that these forecasts already include more than a doubling of coal generation, you can understand why utilities perceive nuclear not as a matter of preference, but, rather, as a crucial necessity.

The estimated shortfall without nuclear energy may take the form of measurable physical shortage, or it might be partially absorbed by the gradual adjustments of the economy to scarce supply that would occur as a chronic condition in the 1980s and 1990s. Because two-third of all electricity is ordinarily consumed by commercial and industrial establishments, the impact on the productive sectors of such shortages would be large, even if they were evenly shared by all users. However, past experience indicates that such shortages are likely to be politically allocated so that the impacts are less on the voting consumers and more on the business establishment. But no matter how the shortages are managed, the impacts on the economy and individual businesses are bound to be severe.

Remember that the electricity consumption estimates with which I began were based upon social expectations of economic growth well below trends of the recent decades. In addition, the estimates already assume a very high level of conservation in the years ahead in response to higher energy prices and improved energy-use technologies.

So these are our real alternatives: more nuclear power, or expanded importation of oil for electricity production, or not enough electricity to meet minimal social needs, even with extensive growth of coal plants. If electricity is not provided, if industry is forced to cut back energy use beyond cost-effective conservation, my concern is that a painful accommodation will be made at the expense of economic output and productive efficiency, or by moving industries to regions where electricity is readily available. Some industries are already facing this situation as a result of foreseeable regional electricity supply constraints. This dilemma generally cannot be avoided by turning to on-site industrial self-generation, because such alternatives are either very costly or require fossil fuels, usually oil. Only a few specific regions have local resources such as wood waste, which can provide an alternate for limited use.

Returning to the theme with which this paper started, under the best of circumstances, the energy outlook for the 1980s must be considered bleak. However, if we can finally rise to this challenge squarely and objectively, we may be able to use the period of the 1980s to implement our technical options to ensure energy abundance in the long run. I believe that the industrial nations must develop and expand those fuel and electricity sources that technology has brought within the range of economic feasibility. And, of course, we must explore new concepts to ensure energy for the distant future. Energy abundance is a foreseeable goal, and at costs which should be acceptable to our economies.

We should cease abetting our public opinion makers in the dangerous platitude that we must face a future of limited expectations because of energy shortages. Limited expectations lead to policies of retrenchment and cautiously riskless investments, thus making the policy self-fulfilling. And, perhaps more importantly, limited expectations destroy the image of a better future so essential to motivate our social institutions.

The basic issue remains our dedication to using all the resources of technology to provide ourselves with energy abundance. The industrial world needs to marshal its available technical resources to develop both energy supply and efficient end-use devices. The scale of living in industrial nations does not have to decline. Higher cost energy supply does not mean a depressed economy, if we permit technology to compensate. We must reestablish our faith in technological progress, and provide the societal freedoms which will permit it to flourish.

PART II

Can Governments Respond in Time?

Chapter 3

International Comparison of Governmental Responses

John C. Sawhill*

I would like to review U.S. energy policy in particular, and briefly comment on steps the industrialized countries can take, and are taking, to meet the challenges posed by the tight oil supply and demand conditions forecast for the rest of the decade.

Throughout the decade we will face continuing pressure on the world oil market. History alone suggests that we should be prepared to face a major supply interruption at least once, and possible twice, during the decade.

The International Energy Agency (IEA) recently published a review of member countries' energy programs in which an anticipated "notional" shortfall of about 4 million barrels a day by 1985 and 8 million barrels a day by 1990 was projected based on policy measures now in place. Using "sensitivity analysis" in which lower economic growth rates and higher prices were assumed still resulted in a "gap" of 2 million barrels a day in 1985 and 6 million barrels a day by 1990. We generally agree with this overall assessment.

The inescapable fact is that we are likely to see a gradual decline in the availability of oil to the West. That trend will be exacerbated by the desire of some members of OPEC to limit production to levels

*Deputy Secretary, U.S. Department of Energy, Washington, D.C.

necessary to meet internal revenue requirements rather than world oil demand, and by the prospects that the Soviet Bloc will become a net importer on the scale of one million barrels a day by the mid-1980s.

We cannot expect that trend to be reversed by increases in production elsewhere. For example, North Sea production will only serve to offset declines in older fields in the United States and Canada. Even the promising finds along the east coast of the United States and Canada still need to be evaluated and will take years to develop. Mexico, with perhaps the greatest near-term potential for increases, has made it clear that future production will be limited through 1982, the term of the Lopez Portillo Administration, to a range of 2.5 to 2.7 million barrels a day.

As we contemplate the oil supply and demand outlook, we must be mindful also of the political instability of the Persian Gulf region. The turmoil in Iran and the Soviet presence in Afghanistan, in Saudi Arabia's neighbors North and South Yemen, and across the Red Sea in Ethiopia, are heightening concern among all of us about the future reliability of supplies.

Therefore, it is vitally important that government policies for responding to future energy uncertainties have as their underlying goal the protection and maintenance of national security and economic well-being. It must be within the framework of our collective and individual security that national and international policies are developed and implemented.

The importance of coordinated national and international policies becomes clear if we reflect for a moment on the events of last year. The cessation of oil exports from Iran during the early part of the year was largely compensated for by increased production by other countries throughout the year. However, in the acute atmosphere of uncertainty surrounding the cutoff, the industrialized countries scrambled to build stocks and ended up doing so at a rate of 1.2 million barrels a day.

The result was that OPEC was able to increase its prices more than 100 percent, which is shocking testimony on the impact of not having adequate oil reserves and management policies to deal with a major interruption. Clearly, if adequate stocks and stock management policies had been in place, we could have mitigated the increases. It is for that reason we must build oil reserves and strengthen demand restraint.

In the United States, we have established a Strategic Petroleum Reserve (SPR). We stopped accepting contracts for oil for the SPR in March 1979 when the market tightened, but I am convinced of the need for renewing SPR fill on a timely basis after consultations with our allies and friends in the consuming and producing countries.

Current market conditions are such that the SPR fill can be renewed without significant price pressures. Our Congress has recently mandated that we begin filling our SPR at a rate of 100,000 barrels per day in an expeditious fashion using domestic production.

It is essential that we proceed with building reserves and implementing policies for stock management if we are to remove our vulnerability to a repeat of the price run-up experienced last year. This is not something the United States can do alone, however. It must involve all industrialized countries, if we are to avoid a repeat of the scramble for oil that allowed OPEC to double its prices last year.

In addition to coordinating oil stock programs, there are other steps the IEA countries need to take. Fortunately, the 21 IEA members have made significant progress in a number of areas such as:

— Designing and testing of an emergency plan to share oil in the event of an emergency;
— developing an oil market information system;
— implementing a long-term program to reduce dependence on imported oil through joint efforts in conservation, production of conventional energy sources, and research and development; and
— providing a forum for coordinating relations with oil producers and developing countries.

At the IEA Ministerial meeting in Paris last month, IEA nations further committed to reduce the energy use/economic growth elasticity to 0.6 through 1990, as well as reducing oil as a percentage of total energy requirements from 53 percent (1978) to 40 percent by 1990. Ministers agreed to undershoot substantially the previously agreed 1985 oil import targets noting that this could require a reduction of 4 million barrels a day below the previously agreed 1985 group oil import objective.

But I believe the principal accomplishment was the establishment of a system of yardsticks and ceilings to help manage sudden supply interruptions that do not reach the 7 percent threshold necessary for triggering the formal IEA emergency oil sharing system. Such a policy tool is essential to dampen the strong upward price movements experienced in 1979. It complements the oil reserve build-up of individual countries and increases flexibility for coping with shortages.

The efforts of the IEA countries will be given additional political impetus at the Venice Summit now taking place. Greater international emphasis on long-term supply alternatives, enhanced conservation efforts, as well as encouragement for the early commercialization of promising energy technologies such as synthetics and renewables, is expected to emerge.

There are many steps individual nations can take to achieve the objectives of the IEA and the Summit. For example:

— allowing oil prices to reflect international levels and natural gas to be priced to encourage its substitution for oil;
— the use of fuels other than oil, and enhanced conservation efforts in the residential/commercial, transportation, industrial, and electric utility sectors;
— increased coal utilization, production and world trade;
— increased oil and gas exploration and development, including the use of enhanced recovery techniques;
— efforts to accomplish projected nuclear programs; and
— government action to accelerate the development and commercialization of new and promising energy technologies.

In the United States energy programs consistent with those actions are moving ahead rapidly. Our primary objectives are to increase domestic production of a variety of energy resources while reducing the use of energy, particularly oil. We have established a goal of reducing our level of oil imports to 4.5 million barrels a day by 1990. While this will improve our national security posture at home, it also will help our allies both by removing some of the competitive pressure that could otherwise push prices still higher and by assuring additional supplies to meet world needs.

We are seeing good progress in reducing demand, which is reflected in oil imports that are down to about 6.7 million barrels a day for the first five months of this year. That is a decline of over 12 percent from a year ago and a drop of about two million barrels a day since 1977.

Oil consumption in 1979 dropped 2.4 percent below 1978 levels and gasoline demand declined 5 percent. This trend has continued into 1980 with gasoline demand down 8.4 percent and total oil consumption down over 9 percent from last year's level. Total energy consumption in 1979 was 0.3 percent below 1978 while the economy grew in real terms by 2.3 percent. That is a significant change from the early 1970s.

Oil and gas price decontrol is now underway, which will further reduce demand while encouraging production. The decision last year to decontrol crude oil prices by September 1981—at a rate of 4.6 percent per month since January—has freed almost 40 percent of

U.S. oil production from controls. The price of U.S. decontrolled oil to domestic refineries is now approaching 90 percent of the world price. By year-end, domestic crude prices are expected to reach $30 a barrel compared to $13 a barrel last year.

The oil industry's response to decontrol has been very positive. More exploration rigs (2,800) are now drilling in the United States than at any other time in the last twenty-four years. As a result of higher prices, the oil and gas industry has increased the number of seismic exploration crews nearly 35 per cent in the last year.

To stimulate production of natural gas we have taken steps toward replacement cost pricing by allowing the phased decontrol of new natural gas production. We have already deregulated the price of natural gas pumped from a depth greater than 15,000 feet, as well as unconventional gas from geopressurized brine, coal seams and Devonian Shales. Our regulatory authorities have approved higher prices for gas from tight-sand formations and are considering additional price stimuli.

Decontrol is just the beginning. We are also developing the technology to make recovery from tar sands and from light and heavy oil reservoirs economically feasible on a larger scale. We now recover only 32 percent of the oil in place.

Expanded use of coal is another key aspect of U.S. energy policy, which we are promoting in several ways:

— A $10 billion commitment over a ten-year period to reduce utility oil and gas consumption by the equivalent of one million barrels of oil a day, replacing it with coal and other fuels.
— An acceleration of coal leasing on Federal land;
— A program to increase coal exports to the 80 to 100 million ton level by the end of the decade.
— Rapidly growing investment in coal R&D, which will top one billion dollars in 1981.

In addition, the indirect use of coal will be helped by a synthetic fuels industry that is already beginning to take shape. Major demonstration projects are well along. To provide incentives we will shortly create a synthetic fuels corporation with initial funding of $20 billion for loan guarantees and price-purchase commitments. Construction of synfuel plants and other high-priority projects will be accelerated by an Energy Mobilization Board, empowered to cut through the regulatory delays which have held up many vital projects in the past. The Administration's goal set for synthetics production is 0.5 million barrels a day oil equivalent by 1987 and 2.0 million barrels a day by 1992.

Nuclear power continues to be an important part of our energy program. Nuclear power plants now provide about 12 percent of our electricity. Plants operating today are contributing the equivalent of 1.5 million barrels of oil per day and by 1985 that number could increase by another 1.5 million barrels.

We are taking a number of steps to insure that the nuclear option remains open.

— First, we are strengthening our efforts to assure that nuclear power plants are built and operated safely.
— Second, we have a program to assure that nuclear waste materials produced by the military and by the private sector are handled and disposed of safely.
— Third, we are working with other nations to find ways of tightening nuclear safeguards so as to limit the risk of nuclear weapons proliferation.
— Fourth, we are improving the regulatory process and expediting nuclear plant licensing and siting.
— Fifth, we have a research and development program that will build a strong technical base for existing light water reactors and for future breeder reactors.

We have also set an ambitious target of meeting 20 percent of our energy needs from renewable resources within 20 years. By the end of 1981 alcohol fuels could replace almost 10 percent of the demand for unleaded gasoline. Our solar program is supported by a research and development budget of nearly $800 million, more than twice what it was just four years ago.

Energy conservation remains one of our highest priorities. In the transportation sector, we have established mandatory fuel efficiency standards for automobiles; fixed a 55 mile-per-hour speed limit for the nation's highways; and undertaken a nationwide program to expand van-pooling and car-pooling; and improve driving habits.

In the buildings sector, we are establishing energy performance standards for new buildings; providing funds to weatherize homes of low-income families; preparing to spend billions of dollars to make homes energy-efficient; and continuing to restrict thermostat settings in public buildings. We also are setting energy-efficiency standards for new appliances. To further encourage both conservation and solar energy in residential and commercial buildings, we will soon establish a solar and conservation bank to provide low-interest loans.

In industry, economic incentives for energy conservation are working well. Since 1972 the ten most energy-intensive industries in the United States have boosted energy efficiency 14 percent—that is equal to eliminating 1.2 million barrels of oil a day.

So I believe the United States is well along in achieving its energy goals and fulfilling its commitments to the world community both as a major producer and user of energy. We know that excessive oil import dependence must be addressed in cooperation with friends and allies, including our friends in OPEC, who ultimately have as large a stake as we do in the evolution of an orderly world. It is essential that we steer a course toward ever-closer cooperation with other industrialized countries to stabilize the global oil market and to assist the less developed countries fulfill their social and economic aspirations. We look forward to the joint strengthening of our economic, political and national security postures through such cooperation.

Chapter 4

General Observations Regarding Governmental Response to Energy Problems and Issues

Ali A. Attiga*

In turning to the subject of this session I wish to stress that my observations are mainly concerned with past developments. I do not claim to foresee the future, but I will submit for your consideration certain suggestions concerning the conditions for international cooperation in the energy field.

I. HISTORICAL PROSPECTIVES

During the 1950s and 1960s the international oil companies were predominant in shaping world energy policy. Governments of oil importing and oil exporting countries responded to the policies of the oil companies in different ways, but generally they followed two main patterns:

1. The oil importing groups responded by encouraging the shift from non-oil energy resources to oil imported from poor developing countries, especially in the Middle East and North Africa. Because of cheap oil imports governments of consuming

*Secretary General Organisation of Arab Petroleum Exporting Countries (OAPEC)

31

countries responded by collecting substantial public funds from taxation on oil products. They generally used these funds for investments in the public sector during the post-World War II period.

2. Governments of the oil exporting countries were for a long-time practically helpless in responding to the cheap crude oil policy followed by the governments and oil companies of the consuming countries.

There were many studies and lectures by learned professionals, academicians as well as governments and company officials of the consuming countries generally stating that there was too much oil and gas in and outside the Middle East and North Africa. We were often told that the rapid development of alternative sources of energy and the abundance of oil reserves will tend to reduce oil prices in the 1970s and the 1980s.

Faced with a rather dismal outlook the oil exporting developing countries' natural response was to put more pressure on the oil companies for more production in order to increase their meager revenues based on average earning of less than 90 cents/barrel during the late 1960s and much less than that for several countries. Prior to 1965 they generally received less than 70 cents per barrel.

World inflation progressively reduced the purchasing power of their export earning with the result that they hardly managed to meet current expenditure needs while more of their investment projects were left with little or no financial resources.

The oil exporting countries responded to their situation on individual basis and generally without much success. Prior to 1965, their only collective action during that period was the establishment of the now famous OPEC, which was either ignored or openly belittled by consuming countries and their oil companies.

Oil-deficit developing countries did not show much concern because their oil imports were small and their energy supply and distribution systems were either owned and/or managed by the international oil companies.

In retrospect the policies of the oil companies and governmental response to these policies prior to 1973/74 led to at least three major results in the energy and economic development fields.

1. Great success in discovering prolific oil fields in the Middle East and North Africa during the 1950s and 1960s.
2. Rapid and wide neglect and general deterioration of non-oil energy resources in the advanced and developing countries alike.

3. Rapid socio-economic development and growth in the oil importing developing countries at a time when there was little or no real economic development in most of the oil exporting and other developing countries.

DEVELOPMENTS DURING THE 1970s

Beginning with late 1960s and early 1970s some oil exporting countries began to demand higher oil prices and when oil companies refused to recognize the need for such demands some OPEC countries, starting with Libya, responded by taking unilateral action to restrict production or take other measures. All this led directly into the unilateral OPEC price increase in October 1973/74.

The initial reaction of the governments in the oil importing industrial countries was vehement and openly advocated hostile action pointing to confrontation measures designed to break-up OPEC and consolidate the bargaining position of oil consumers. First, there was an attempt to draw oil exporting countries into bargaining the price and supply in isolation from other developing countries (note the Washington Energy Conferences in February 1974). Then there was the attempt to draw non-oil developing countries in the same group with the developed countries in facing OPEC countries. This was clearly evident during the so-called North-South Dialogue held in Paris 1975–77. While both attempts have not so far succeeded in achieving their objectives there is much evidence that they have not been abandoned by their advocates. From 1973 to the present the OPEC countries have generally followed a policy of pricing their oil exports on the basis of a series of *ad-hoc* responses and compromises based on the economic policies of the advanced oil importing countries. They have been following rather than leading general world inflation, currency fluctuation and oil market conditions, all of which fall outside their control and jurisdiction.

THE ENERGY OUTLOOK FOR THE 1980s–1990s

There are great risks and equally great opportunities depending on whether government actions and policies responses of oil exporting and importing countries take the form of confrontation or cooperation.

The OPEC countries face the great risk of seeing their major, and in many cases the only, natural resource rapidly depleted before they can create alternative sources of income and employment. The oil importing countries may face the risk of severe shortage of oil supplies before they can develop alternative sources of energy.

But it is possible to argue that these great risks can be avoided and even turned into new opportunities for cooperation in the development of global energy resources and in the allocation and use of these resources. For success in this promising possibility it is necessary, I think, for the oil importing advanced countries to accept as legitimate, necessary and even desirable at least four main conditions:

1. Progressive increase in the real price of oil with declining proportion of supplies from OPEC countries. OPEC countries have less than 9% of world energy resources, but provide more than 16% of world energy supplies. They should be expected to continue to carry such a heavy burden on their limited energy resources.
2. Increasing dependence on non-OPEC oil and other sources of conventional and new forms of energy.
3. Increasing exports of oil and non-oil energy resources to the energy deficit developing countries with due regard to their economic and financial needs where appropriate. OPEC countries, in spite of the fact that they are all developing countries generally depending on depletable resources, have been providing considerable assistance to other developing countries. The same can not be said of the rich countries of the OECD or the Comecon countries.
4. Increasing technical and financial assistance to developing countries for the discovery and exploitation of their domestic energy resources. The advanced industrial countries should provide more of this form of assistance because they possess the necessary technology and resources.

It is highly possible, in my opinion, that if these conditions can be accepted by the powerful and wealthy industrial countries then there may be a real chance for a sound and balanced international approach to the international energy issue of today. Regional and individual governmental policies and responses will then have a general framework for global energy transitions from oil to non oil resources over a reasonably long period of time, which should give the different parties chance to accommodate their different needs without pushing each other for short-term advantages.

Chapter 5

A U.K. Government View

I shall start by outlining what my own Department's percep-
tions of trends in the next ten years look like.

(a)   we estimate demand for OPEC oil this year at about 28 mbd;
and supply of OPEC oil at or slightly above this figure. In
other words, a balanced market, in the sense that there
should be no general pressure on supplies. But this does not
necessarily mean stable prices—two rounds of "leapfrogging"
as a result of Saudi attempts to re-unify the OPEC pricing
structure have taken place since the beginning of the year;

(b)   if we postulate a low economic growth rate (2%/year) and a
small real increase in oil prices (also 2%/year), by 1985
demand for OPEC oil is likely to be around 27 mbd and by
1990 around 28.5 mbd. A lot of political guesswork has to
go into projections of the supply of OPEC oil—but with a
working assumption of OPEC production in the range 28-30
mbd, it is possible—on a 2% GNP growth assumption—that
there could still be a balanced market in 1985. But the
previous caveat about prices still applies. And by the end of

*Parliamentary Under-Secretary of State, U.K. Department of Energy

the decade, demand for OPEC oil will be within the likely range of willing OPEC supply and therefore vulnerable to production shortfalls;

(c) if we assume that the Western economies are capable of growing faster than 2%/year (we have used as a high case 2.5%/year to 1985 and 3.3%/year thereafter), then the picture becomes much more gloomy. Demand for OPEC oil in 1985 would be around 31 mbd and by 1990 above 35 mbd—both figures above a working assumption of 28—30mbd for willing OPEC supply.

All these figures—and the detailed figures lying behind them—are full of uncertainties, uncertainties magnified by the fact that demand for OPEC oil is residual, after all other sources of oil and non-oil energy have been accounted for.

Governments need, however, to draw conclusions and develop policies, even when the projections are very uncertain. They need to examine and try to assess the upside possibilities and the downside risks.

The upside possibilities clearly do exist and are worth stressing, given the almost inevitably sombre view of the next decade which most informed commentators now embrace. OPEC production in 1990 could be significantly higher than the present level. This is certainly physically possible but it depends upon OPEC countries (particularly the Gulf States) taking political decisions to undertake the necessary investment at a fairly early date. Second, our projections and those of others may under-estimate the *long-term* impact on consumer habits of last year's doubling of oil prices. We only have historical experience to work on and that is not an infallible guide. The marked fall in US oil consumption in the last 12 months is an encouraging sign.

So there is a glimmer of hope. But there are also some very large downside risks, which almost certainly outweigh them. There are the political risks of disruption to supply—even a small disruption in volume terms could have a large impact on market psychology and hence on prices. There are the risks that we and others have over-estimated the availability of non-OPEC oil (eg in the United States) or the speed at which non-oil energy can back oil out of markets. There is the risk of lower OPEC production. A number of OPEC countries are currently generating more revenue than they need for their immediate purposes and may do so for much of the next decade. They may conclude that their interests are best served by leaving oil in the ground. Others want to spin out their limited reserves.

The conclusions that governments must draw are that there is a very real risk that the 1980s will see one or more periods of tight oil market conditions accompanied by rapid rises in oil prices. The timing of these periods and the extent of possible oil price rises are extremely uncertain, however.

It is important to realise the wider economic implications of this conclusion. Oil is by far the most important single commodity in world trade. Sudden movements in prices shift large resources from consumers to producers and reduce demand in Western economies to the extent that OPEC countries cannot or do not choose to spend their incremental revenue within the OECD area. As a rule of thumb, a $2/barrel rise in average oil prices:

(a) reduces OECD GNP by 0.3% after 1 year
(b) increases inflation by 0.5% after 1 year
(c) results in a $17 billion deterioration in the OECD current account position in a full year.

The OECD Secretariat have calculated that by 1981 the real income loss to the OECD from last year's doubling of oil prices could be $300-$400 billion—or $1,500-$2,000 per family. The U.K., as a major trading nation, cannot be shielded by North Sea oil from adverse affects on the world economy.

The implications for developing countries are serious too. The current account deficit of non-oil LDCs in 1980 will be double the 1978 level (even assuming no further increases in the price of oil). As they become less able to sustain deficits of this order, their economic growth will be cut back. The poorest LDCs, who depend heavily on aid flows, will have a struggle even to stay where they are now.

Analysis suggests the possibility of an oil constraint to growth in the 1980s. If Western economies grow fast, thereby increasing the demand for oil—there will be pressure on oil supplies and a sharp increase in prices with a marked fall in economic growth. This is not the place to speculate on the growth rates that might be possible without coming up against an oil constraint; but it is worth recalling the figures I mentioned earlier—a 2.5% growth rate to 1985 and 3.3% from 1985-90 (ie very modest growth rates compared with the 1950s and the 1960s) seems likely to put severe strain on the oil market before the end of the decade.

It is natural, therefore, that energy prospects should have run like a thread through so much of the discussion at the Venice Economic Summit on 22/23 June 1980. Summit leaders focussed on medium-term energy problems (ie to 1990 roughly) in the context of their economic significance.

The Venice communique began with the words "the economic issues that have dominated our thoughts are the price and supply of energy and the implications for inflation and the level of economic activity in our countries and for the world as a whole. Unless we can deal with the problems of energy, we cannot cope with other problems". It continued by noting the enormous economic damage which had been done to both the industrialised countries and the developing world by successive large increases in the price of oil.

The Summit mapped out in broad outline the course the industrialised countries will have to follow in this decade and beyond. It made it clear that the key task was to break the existing link between economic growth and the consumption of oil; and in this context stated the firm belief of Summit governments that maximum reliance had to be placed on the price mechanism and that domestic oil prices had to take world prices into account. The Summit communiqué was also quite clear about the policy measures that will be needed—no new oil-fired power stations should be built, except in exceptional circumstances; there must be increased efforts to substitute for oil in industry and the commercial and residential sectors; the introduction of more fuel-efficient vehicles must be accelerated; and above all there must be a large increase in the use of coal and nuclear power in the medium term and in synthetic fuels and the renewables in the longer term.

The Summit went beyond mere verbal descriptions of what was necessary, to specify in figures what the Summit countries as a group aim to achieve. The Summit countries stated their intention of making a co-ordinated effort to increase the supply and use of non-oil energy by 15–20 mbd over the next 10 years; coal production in the Summit countries as a group should double by 1990; the ratio between increases in energy consumption and economic growth should fall to 0.6 (from about 1.0 at present) and the share of oil in total energy demand should fall to 40% by 1990 (from 53% now).

These are bold and ambitious targets. Some cynics might suggest that governments only agree to international targets when they are fairly certain they can be achieved with only minimal effort on their part. This is emphatically not the case when it comes to this Summit's commitments. Individual countries will have to demonstrate their will to achieve these targets in the manner most suited to their national circumstances. And determination is only half the story. Each country will have to develop the means to achieve the desired end. This Summit declaration is a firm affirmation that the Western nations will not duck this issue: the Venice commitments will require vigorous and decisive policies by the governments of all industrialised countries. The doubling of coal production in the Summit countries will depend critically on the expansion of the U.K. coal industry and a rapid increase in the international coal trade. This in turn means substantial investments in coal handling and transport facilities and in coal-using equipment in the consumer countries.

The contribution of nuclear energy, which at present accounts for only 3% of the Free World's energy needs, will have to expand substantially if the Venice Summit goals are to be realised. The Summit communique stressed the vital contribution of nuclear power to a more secure energy supply. A recent Shell study suggested that it could be possible to raise the nuclear contribution to 10% by 2000; this would require the construction of 500 nuclear power stations in the non-Communist world, at a cost of more than $600 billion in 1980 dollars. I cannot speak for the figures, but they are a clear indication of the magnitude of the task which faces us.

The Summit leaders did not focus exclusively on the energy problems of the industrialised countries. They also recognised that a major international effort was needed to help the developing countries increase their energy production. The World Bank is already playing an active role in this area. Its programme of lending to developing countries for oil and gas exploration and development will expand to over $1¼ billion in 1983 and a smaller coal programme will expand to $340 million by the same year. An important feature of the Bank's programme is the extent to which it hopes to attract counterpart investment from other sources, particularly the private sector. But this effort may not be enough and the Venice Summit therefore asked the World Bank to consider means—including the possibility of establishing a new affiliate or facility—by which it might improve and increase its lending programmes for energy assistance. The Summit asked that the Bank should explore its findings with both oil-exporting and industrial countries.

The Summit touched, too, on the important question of producer-consumer relations. It restated the West's interest in a constructive dialogue on energy and related issues between energy producers and consumers. The Government believes that this is a policy area of major importance. If producer/consumer relations go badly wrong, the consequences could be disastrous. Equally, if by intelligent policies we succeed in creating a better understanding between the two groups, we should be able to enhance the perception or common interest between them and measurably increase the chances of maintaining reasonable levels of economic growth through the next decade.

The oil producers have a powerful interest in the economic health of the industrialised world—their investments in Western economies, their interest in maintaining stability in the prices of the industrial products they import, the extent to which good economic performance in the West helps them to diversify their own economies are all considerations which underline the fact of inter-dependence.

I think that a developing relationship between producers and consumers is not only desirable but distinctly possible—a relationship based on a measure of understanding about the nature of the problems of immediate interest to both sides and which, by virtue of a joint consideration of the world's energy outlook, would contribute to future stability in the oil market. Such a relationship, reflecting a new understanding of the inter-dependence between oil producers and consumers would require changed attitudes on both sides.

Consumers need to continue to demonstrate their recognition that conventional oil is in limited supply, that the general trend of real oil prices is likely to be upwards and that it is not unreasonable for producers to develop depletion policies which reflect their concern with future generations and the need to ease the process of adjustment to an oil-scarce future. Above all consumers need to show that they are serious about the re-structuring of their energy economies and reducing their dependence on oil. In this context the activities of the IEA—particularly its longer term work—are constructive rather than confrontational.

The attitudes of producers are just as important. There is a continuing need to recognise the wider economic repercussions of decisions they take on oil prices—not least on developing countries. And just because oil prices and supplies are difficult and delicate areas does not mean that they should not be discussed.

In conclusion, we face a difficult decade. Energy problems will be one of the key areas. There are no easy solutions. The long-term answer is to diversify away from oil and to use energy more efficiently. The adjustment will take time—major energy projects typically have a 10 year lead time. Oil is not easily substitutable in many of its uses (eg transport). In industry a move away from oil means large investments in new plant (not necessarily easy if at the same time general economic prospects are poor). It is necessary to avoid trying to shield consumers from oil price increases—the market mechanism is essential to adjustment. It would be disastrous to try to inflate out of trouble, to award ourselves wage increases to pay for the higher cost of imported oil. It is essential to look again at relations with the oil producers; OPEC is not a band of brigands holding the world to ransom. It is a group of countries motivated, as we are, by a perception of their long-term economic interests. We must recognise this and recognise the potential stabilising role that OPEC could have in the oil market. We must look hard at the possibilities for developing a new relationship between consumers and producers, which can be of benefit to both and to the developing countries.

PART III

Energy in World Power Politics

Chapter 6

Key International Energy Problems of the 1980s

Melvin Conant*

The key energy problem of the 1980s may be summed up in the single question: How will oil importers obtain their supply? The general anticipation is still one of tightness in supply and quite possibly contrived shortages. The competition between us will be corrosive of all our other relationships. And how may the producers react to our respective and divisive efforts?

Will NATO nations and Japan be able to agree on a set of common strategies and agree on their expectations of each nation's role in the obtaining of supply or, if defined, will these be observed? Will the security alliances (NATO, Japan-United States), including whatever may be attempted in the Gulf, be gravely impaired as a consequence of competition for supply? If so—and it seems likely—how will this affect the present strategic imbalance between NATO and the U.S.S.R.? It will scarcely be improved and may be greatly worsened. And the oil producers will be implicated in what ensues.

Energy—access to oil—is not the only factor which could weaken our common and particular defense (political will, economic relations, defense budgets, etc.) but it is now to be ranked among the very highest interests of state and therefore to be found in so much

*Conant Associates

of what we do. This relationship between energy and a whole range of interests is not yet adequately understood, least of all perhaps in the United States; energy is still considered in too narrow a context.

What should worry us most?

1. A European initiative towards Middle East producers in which neither U.S. nor Japanese oil needs will be considered. The implications of such an initiative with its warning of exclusivity would most profoundly affect us all. (As would too great a dependence upon gas exports from the U.S.S.R.)

2. An attempt by the United States to fashion a hemispheric energy system—admittedly in the longer range— in which neither European nor Japanese oil needs for diversification of supply will be considered, nor the particular perspectives of Venezuela, Mexico or Canada be respected. Again, the implications of such an initiative could seriously impair our alliances.

3. The "isolation" of Japan. There is a paradox here; Japan is an economic giant in world trade, deeply engaged in virtually all industrial and developing markets. Its energy needs are massive (it is the second largest importer) and its continuous supply is a constant concern. Yet, despite links to the "western" international financial and economic system it is still not considered by us—or by them—to be fully integrated into it. The political separateness which still characterizes their nation is a standing reminder of how necessary it is that the nation's fullest incorporation into the international system be evident to all. Unless this is accomplished, Japan may believe itself compelled to go its own way—again—and the geopolitics of Northeast Asia, including the PRC, will once more haunt us.

4. The failure to have a meaningful dialogue between the oil producers and importers. As it is, the dialogue is one of the deaf, as it has been described. Neither side truly comprehends what each is doing to the other; if there is a sense of urgency to make progress, it seems to be almost wholly with certain key importers. Why should it still be so difficult for both sets of interests to agree we have a common stake—if for somewhat different reasons—in the prolonging of the oil era? Each side needs time to prepare or "manage" its affairs for the inescapable, eventual transition from oil.

5. A failure on the part of the industrial world—especially the United States—with the full concurrence of key Gulf producers, to make clear and convincing to the U.S.S.R. that its oil needs

can be met within the "system" for the obtaining of oil, if the U.S.S.R. elects to take that course.

6. A failure on the part of the United States to realize upon its many energy options, in time (to lessen its dependence upon imports and to diminish the competition for oil which is anticipated to be in short supply).

7. A failure on the part of the United States—in concert with Saudi Arabia—to enlarge upon the Kingdom's political, economic and security relations with Europe and Japan. A lower U.S. presence in Saudi affairs is the objective. It would be the height of foolishness for that "paramount" American influence to be so singularly expressed when the winds of political change continue to blow across the region.

Every one of these points underscores the "politics" in access to oil, a reminder to energy economists that their discipline is a necessary but not sufficient tool to cope with the realities of supply.

Chapter 7

East-West Factors in International Energy Production and Trade

*M.C. Kaser**

Professor Odell has just suggested that, contrary to much emphasis on conservation and on alternatives to oil in production, the key to the Western energy problem lies in exploring for and producing more oil.[1] To some extent that is true for the East—the Comecon group of the U.S.S.R., East Europe, Cuba, Mongolia and Viet Nam. China, too, should enter our consideration, but for the present brief exposition I feel justified in putting in only a single mention of the Asian countries with ruling Communist parties. It is, like Professor Odell's prognosis for the Third World as a whole, potentially energy self-sufficient for much higher levels of industrialization. The large and dispersed petroliferous areas, inland and offshore, in China and Viet Nam and the substantial coal deposits of China and Mongolia will support even the one billion population of that group at feasible requirements for the rest of the century.

*St. Antony's College, Oxford

[1] All quantities are in barrels of oil equivalent. Soviet and East European data are, however, invariably cited in tonnes of coal equivalent; the following (approximate) conversion factors were used: 1 tonne of oil equivalent (10,000 kcal/kg)=1.43 tonnes of coal equivalent (7,000 kcal/kg) and 1 tonne/year =50 barrels/day (b/d).

With 20 million additions to the workforce annually there is no constraint on labour, but capital and technology transfer will constrain the speed of expansion of energy-using production.

That said, I want to concentrate on the energy surplus of the U.S.S.R. and the possibility—forecast by the United States Central Intelligence Agency—of a shift early this decade into balance and then into a net import deficit. I shall first discuss the U.S.S.R. and then Eastern Europe as a group; the latter excludes Albania and Yugoslavia, respectively a net exporter and a net importer.

One is helped by the publication last year of an official Soviet energy balance. It differed from the United Nations estimates by showing slightly higher production and slightly lower consumption; the margin of interest to the rest of the world was thus a little more generous. The Soviet figures run from 1913 to 1978 and, to take a recent period, show that the Soviet excess of production over consumption increased from 1.9 million b/d in 1965 to 3.7 million b/d in 1978. After allowance for changes in inventories (a rise in the year 1965 and a fall in 1978), net exports rose from 1.5 to 3.8 million b/d. Gross exports in 1978 were 4.3 mn b/d—oil and gas to both East and West Europe and small deliveries of electric power to East Europe. Against that, the U.S.S.R. bought gas from Iran and Afghanistan and a little oil (some of it delivered on Soviet account without actual transhipment in the U.S.S.R.).

The last year for which we have, as yet, official Soviet data on aggregate energy output and use is 1978, but even for that year trade must be estimated because it was then that the U.S.S.R. suppressed tonnages in its trade returns. Production statistics remain published: in 1979 oil output reached 11.7 mn b/d (against 11.4 mn b/d in 1978) and is planned to be 12.1 mn b/d this year. This is significantly below the Five-year Plan target for 1980, as established at the end of 1976, namely 12.8 mn b/d and I say immediately that that is the goal I expect to be formulated for 1985 when a new Five-year Plan is approved by a Party Congress next year. Actual output, based on returns for the first four months, suggest that this year's production will be a trifle under target, at 12.0 mn b/d. In short, a levelling off of production in contrast to the rapid rise under the present Five year Plan—23 percent even with the expected shortfall from the target. Within those dynamic five years, the share of the old areas (chiefly the Caucasus and the Volga-Urals deposits), fell (on 1975 to 1979 data) from 66 to 47 percent and among the new areas West Siberia alone rose from 30 to 48 percent.

It was chiefly on the postulate of a peaking out of the huge Samotlor and other West Siberian producers that the CIA forecast a Soviet output of as little as 8 mn b/d by 1985. The CIA rightly noted (as Soviet experts have since acknowledged) that the major West Siberian fields were experiencing serious water encroachment because a water injection technique was employed which maximized initial output per well; although this minimized investment per unit of output at the early phase, the share of water in total fluid lifted could not fail to rise. From 50 percent or so in 1975, the CIA forecast a share of 65 percent by this year. The large quantities of high-capacity submersible pumps and gas-lift equipment needed to stave off a fall in oil yield were in excess of home production or import possibilities. With a second line of argument that exploration drilling was insufficient to discover new reserves, the Agency's most pessimistic forecast was that by 1985 "the Soviet Union and Eastern Europe will be importers of 3½-4½ million barrels per day".

The CIA has not formally withdrawn its low estimates for Soviet 1985 output, but early last year it took the upper end of its range as more likely (10 mn b/d) and in a more recent study, of August 1979, has avoided an explicit statement on output, concentrating on the likelihood of a reduction of its net oil exports from 3.0 mn b/d in 1978 to 1.7 mn in 1982 and foreseeing insignificant net energy exports by 1985.

The important elements in the selection of a forecast for 1985 exports to the West are the availability of technology embodied in imports from the West, chiefly the United States, and the requirements of Eastern Europe. The evidence at the moment is that the United States Government is willing to sell the appropriate equipment to the U.S.S.R., despite the embargo on high technology which it imposed as a result of the Soviet invasion of Afghanistan, although the knowhow or equipment to manufacture such requirements will not be allowed to be sold. Moreover, Western European firms are unaffected by the embargo. Thus, the U.S.S.R., if it wishes to buy them for hard currency, can obtain the submersible pumps, the offshore equipment, or the rotary drills for deep bores in soft rock that it cannot in the medium term furnish for itself. The only reason why the U.S.S.R. might not spend the $10 billion over the medium term which seems to be required is the United States' embargo on the other products it wants to buy. This is the element of uncertainty in the question: the U.S.S.R. earns over half its hard currency from oil sales but will it invest hard currency to generate those sales if it cannot buy grain and the technology it needs to modernize the rest of its industry?

The other feature which affects the availability of oil to the West is East European demand. Consumption of total energy reached 8.4 mn b/d in 1978, a rise of 58 percent since 1965; by contrast, its production had risen by only 37 percent, to 7 mn b/d. East Europe's deficit has sextupled since 1965 and was in 1978 1.4 mn b/d. Soviet oil supplies more than covered that deficit (1.6 mn b/d in 1978), but they had to offset the sales of Polish coal which Poland urgently needed for hard currency.

Polish needs for hard currency have, due to its heavy indebtedness, since become even more acute.

East Europe's problem is that it must earn hard currency both to buy more oil than the U.S.S.R. will allocate and to replace earnings as it is compelled by rising consumption to forgo some of its coal exports. Where East Europe has an advantage is in having a high share of its present consumption in the solid fuels it has in relative abundance and in having hitherto restrained demand for the liquid fuels it has in short supply. In 1978 65 percent of East European energy consumption was of solid fuels and only 19 per cent in liquid fuels. Soviet pipelined gas had brought gas consumption to 15 per cent. Nuclear and hydro power provided only 0.7 per cent (not much different from the Soviet 1.6 per cent).

The issue for the longer-term future is whether that restraint can continue. As it is, perhaps a quarter of its oil needs will have to be covered from hard-currency sellers (i.e. OPEC) by 1985. Its energy input per unit of GDP is already high. In three East European countries where GDP per head is only 40% of that of the United States, energy intensity is 90% of that of the United States. This is a danger sign as much for the U.S.S.R. as for East Europe, for the U.S.S.R. has a heavier (35 percent) reliance on oil. By 1985, nevertheless, with enough Soviet oil to supply its Comecon partners and some to spare for the West and ample Soviet gas for both East and West Europe, the East will relax, rather than tighten, the energy constraints of the West.

PART IV

Capital Constraints and Opportunities

Chapter 8

Financing Energy Developments in the Third World

*R. K. Pachauri, PhD**

INTRODUCTION

Debates and discussion on energy issues in recent years have focussed largely on the problems of developed countries. Despite a growing realisation of the economic interdependence between developed countries and the LDCs, very little attention has been paid to the future energy problems and prospects of the latter. The Third World is not monolithic in its economic, social or demographic characteristics, and there is now a clear distinction between the problems facing oil exporting developing countries as against those of the net importers of energy. In this paper, we would like to concentrate essentially on the oil importing developing countries (OIDCs), since they represent the worst affected societies in the present energy picture of the world.

The concern for assessing and projecting the problems of the OIDCs, apart from its humanitarian aspects, is justified by clear reasons of mutuality of interests between the developed countries and the OIDCs. Historically, the OIDCs have supplied large quantities of raw materials to the industrialised countries of the West and

*Director, Consulting & Applied Research Division, Administrative Staff College of India

imported from them manufactured goods. This relationship while continuing is likely to expand with the growth of markets in the populous OIDCs. The long run interests of the West, therefore, coincide with the desirability of rapid growth in the LDCs in general and the OIDCs in particular. Any slowing of economic growth in the latter will slow down the export of capital as well as consumer goods from the developed countries, which would adversely affect their own economic activities.

Further, with the rapid advances in communications, awareness of global problems and literacy, the opulence of the industrialised countries would stand out in stark and dangerous contrast with the vast ocean of poverty in the OIDCs. The result would be a heightening of global tensions which may be manifested in situations similar to what has happened recently in Afghanistan. Energy, so necessary for fueling the engines of economic growth is a resource the global scarcity of which today threatens the very survival of a large body of nations.

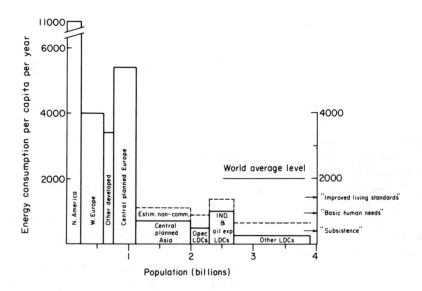

Source: Report of the Policy Analysis Division, Energy Needs, Uses and Resources in Developing Countries, Brookhaven National Laboratory, March 1978.

Figure 8.1. Distribution of world commercial energy consumption. Annual per capita consumption vs. total population (area gives total energy).

The skewed distribution that exists today in the levels of energy consumption is exhibited in Figure 8.1 which shows energy consumption per capita on one axis and population on the other, for different groups of countries.

Not only does it appear impossible to rectify this inequitable form of consumption in the near future but even maintaining these levels is a serious problem created by rapid oil price increases since 1973. It was in response to this problem that the United Nations in 1974 adopted a declaration on the establishment of new international economic order. Whereas the programme of action adopted by the United Nations was broadbased in nature, it arose largely out of the immediate problems of high oil import bills for a number of member nations and addressed itself specifically to means for redressing this acute crisis. Unhappily, over the last six years, very little follow up has taken place both within and outside the United Nations for implementing both the spirit and content of this declaration.

THE IMMEDIATE PROBLEM
OF THE OIDCs

Commentators have often put forward the view that the OIDCs would somehow muddle through for, after all, the prophesies of doomsday and total collapse made in the mid-70s with respect to the most seriously affected countries have not come to pass. This provides hope for the future, but as we would endeavour to establish, does not provide any basis for abandoning the responsibility of the international community in dealing with future problems of the OIDCs. Undoubtedly, in the last 5 or 6 years some healthy developments have taken place in the OIDCs particularly in evolving more rational prices for energy, fostering and promoting exports of various goods, and in developing suitable institutions for energy policy formulation and implementation. Further, particularly in the case of countries such as India, Pakistan and Egypt, the availability of large indigenous skilled labour has been utilised for export of human capital to capital-rich, but labour-deficient countries particularly in the Middle East. The earnings from this export of human capital both in the form of immigrant labour and construction contracts have softened the blow of higher oil prices for some of these countries. There is enough evidence to suggest, however, that the "limits" of this complementarity in foreign trade have now been reached and the future would be much more difficult than the past. Further, the worst cases to be considered are those of nations which have very few raw materials or finished products for export and are heavily dependent on imports for meeting their demands for energy such as countries like Nepal, Bangladesh, Ethiopia and Somalia. The problem in general can really be divided into three components:

a) The immediate future which poses serious problems for import of oil and adverse balance of payments for the OIDCs;

b) The medium term which calls for major investments to develop indigenous supplies of energy and alternative forms of energy consumption, and

c) The long term which requires a transition from conventional to non-conventional sources of energy.

Let us first address the immediate problem in aggregate. The impact of oil price increases on the OIDCs can be appreciated on the basis of figures given in Table 8.1.

Table 8.1. Balance of payments on current accounts[1] 1973-79

	(US $ Billion)						
Group	1973	1974	1975	1976	1977	1978	1979[2]
Industrial countries	19	-4	25	7	4	33	10
More developed primary producing countries	1	-14	-15	-14	-13	-6	-10
Major oil exporting countries	6	68	35	40	32	66	43
Non-oil developing countries	-11	-30	-38	-26	-21	-31	-43
NCL Position[3]	15	20	7	8	2	2	–
Gross Imbalance[4]	(11)	(48)	(53)	(40)	(33)	(37)	(53)

Source: IMF Annual Report, 1979.

[1]On goods, services and private transfers.

[2]As projected by IMF staff on the basis of the price rise of oil decided upon in June 1979. It is estimated that the surplus of the oil exporting countries will be as high as $80 billion in 1980 taking into account the further rise in oil price decided upon in December 1979.

[3]Reflects errors, omissions and asymmetrics.

[4]Yearly totals of group deficits on current account only. The imbalance thus reflected is probably very much understated because the figure for each group is still that of deficit net of surplus, if any, enjoyed by a country within the group.

These figures show a rising trend in the balance of payments deficits of the OIDCs as a group, during the last two years. It is also interesting to note that the initial adverse impact in 1974–75 was partly reversed in 1976–77 when oil markets remained reasonably stable. The prediction for 1980 and beyond is much worse, and some estimates indicate the total deficit is likely to touch $100 billion. This scale of imbalance cannot be sustained without massive relief from the oil exporting countries as well as the major oil consuming countries who in spite of economic difficulties continue to remain in a position of relative advantage.

One important feature of current monetary flows is the trend in recycling of petrodollars which overwhelmingly favours the developed nations particularly in West Europe and North America. One cannot find fault for this development, since investors in the oil rich nations naturally prefer investing their wealth with established financial institutions at minimal risks. At the same time, curbs and restrictive policies by some OIDCs undoubtedly choke off possible flows of recycled dollars to these nations, but there is apparently a case for establishing mutlilateral institutions that could help in reversing the flow of petrodollars at least in part so that the OIDCs would have access to additional means for financing imports in the short term.

There is also a strong case for a dual system of pricing in oil in favour of the OIDCs. In sheer economic terms this would not appear to be efficient in the long run, since undoubtedly prices lower than free market levels would promote sub-optimal use of this depletable resource, and encourage technologies that are inefficient in energy use. The problem, however, is one of phasing the transition with existing technologies and capital stocks. Countries that are making an effort to industrialise would have to slow down their industrialisation plans and wait for new non-energy intensive technologies and capital equipment to resume this effort. The hardship and dislocation of such a strategy would be more serious than the long run benefits of higher and rational energy prices. The dual pricing system that could be adopted may very well fix quantity quotas for eligibility for concessional prices and any excess would require imports at existing market prices. The other fear associated with dual pricing is one of large scale diversion of imports by the importing countries in order to reap profits on the concessional imports. This is unlikely to happen since the compulsions of growth do not allow any reduction in energy consumption in the near future for this type of black-market activity, and further if this is done as an expedient in any case, the importing nations would sooner or later have to import a similar or larger quantity from the world market. Undoubtedly, the system proposed would have difficulties in administering, but for this purpose a dialogue between oil producers, major oil consumers and the OIDCs must be instituted. Most dialogues and debates in the past have been either between the various oil exporting countries or between the major consumers, and the OIDCs have not been included. The future importance of the OIDCs even as consumers is underlined by the fact that the demand for energy in the OIDCs would go up from 12.6 MBDOE in 1976 to 28.5 MBDOE in 1990 (Lambertini, 1979).

MEDIUM AND LONG TERM DEVELOPMENTS

If we were to accept that the international community is interested in promoting global economic welfare, the dynamic relationship between energy demand and GDP must be kept in mind. Historical developments indicate that the energy/GDP elasticity declines with economic growth. Estimates for this measure are 0.81 for the United States, 0.76 for France, and 0.73 for the Federal Republic of Germany. As against this, the elasticity for India was 1.95 between 1953-54 to 1960-61 and 1.86 for the period 1970-71 to 1975-76. In the recent report of the Working Group on Energy Policy, prepared in India, forecasts of India's demand for energy have been produced as shown in Table 8.2.

The energy GDP elasticity implied in these forecasts incidentally, assumes a decline from 1.34 in 1975-82 to 1.08 in 1992-2000. In

Table 8.2. Reference level forecasts of energy demand 1982-2000

	(in million tonnes of Coal replacement)			
Energy from	*1982-83*	*1987-88*	*1922-93*	*2000-01*
Coal*	96.8	131.5	186.6	308.0
Oil*	165.1	217.1	290.6	482.3
Electricity	128.3	191.2	282.0	471.0
Total commercial				
energy	390.2	539.8	759.2	1261.3

Source: Report of the Working Group on Energy Policy, Government of India, 1980.
*Includes only quantities used directly for energy uses.

spite of this overall improvement in energy efficiency, the task facing a poor nation such as India is obviously staggering. In effect, coal consumption is expected to go up to three times the present level, oil to almost three times and electricity to 4 times its present level. The scale of these developments must be seen in the context of maintaining present levels of economic activity by importing adequate quantities of energy in the immediate future. India's oil import bill is already in excess of 50 per cent of its export earnings.

Compounding the simultaneous problems of importing and developing supplies of commercial energy is the widespread use of non-commercial forms of energy in a large number of OIDCs. The consumption of non-commercial energy in India for instance is approximately 44 per cent of the total, but in some countries in Africa it is as high as 85 to 90 per cent. The social costs of non-commercial energy consumption can be enormous, and in a number of cases large forest resources have been depleted with adverse environmental effects. Increased incidence of floods and silting of rivers is a direct consequence of deforestation in the catchment areas of major river systems. Another adverse feature of large scale non-commercial energy consumption is the thermally inefficient manner in which energy is consumed. It is not unusual to find only 5 per cent of the heat content available in non-commercial sources of energy in actual cases is being transferred usefully in the process of consumption. Since the private cost of non-commercial energy is generally limited to the opportunity cost of collection and consumption, optimal technologies are not developed and utilised by consumers, even though with greater sparseness in forest resources around human habitations, the opportunity cost of firewood and animal dung collection has been increasing rapidly.

Unfortunately, the non-existence of organized markets for energy supply in rural areas of most OIDCs does not promote the use of alternate sources of energy. Even if markets did exist, low income levels in these countries would not lead to the consumption of commercial forms of energy. The result is that in spite of some economic growth in the aggregate, non-commercial energy consumption has not decreased in absolute terms in recent years. This is an area in which national governments themselves have not done enough, but there is a case for international assistance whereby effective planning methodologies and the installation of decentralised systems for energy production could be considered and provided by the industrial countries. This presupposes, of course, that the industrial countries themselves develop expertise and capabilities in these areas. Another possibility which has considerable potential is the transfer of skills between the more advanced OIDCs to those who do not possess the requisite skills.

The sum and substance of any analysis of the medium term is that the OIDCs will continue to demand much larger quantities of energy in the future to ensure economic development, and increases in income will lead to some substitution from non-commercial to commercial sources but this will not reduce sufficiently the aggregate energy intensity to bring about a significant departure from existing forecasts. Table 8.3 shows recent trends in economic growth and energy consumption which clearly indicate that even though the oil price increases of the mid 1970s have had some impact on energy intensity in the developing countries, growth in energy consumption is still ahead of economic growth.

Table 8.3. Economic growth and energy consumption, 1973–78

	Economic Growth Rates (% per year)		Commercial Energy Consumption Growth (% per year)		Change in Energy Intensity (% per year)	
	1970–73	*1973–78*	*1970–73*	*1973–78*	*1970–73*	*1973–78*
World	5.6	3.4	4.8	2.6	–0.8	–0.8
All industrial countries[a]	5.4	2.9	4.0	1.6	–1.3	–1.3
All developing countries[b]	6.7	5.3	8.4	6.4	1.7	1.1

Source: Gordon, Lincoln; Energy and Development: Crisis and Transition. Unpublished paper, 1980.

[a]Eastern and Western Europe (excluding Greece, Spain, Portugal, Yugoslavia), Australia, Canada, United States, Japan, New Zealand.

[b]All others (including oil producers, People's Republic of China).

FINANCIAL AND INSTITUTIONAL CASES

Various estimates of the energy related financial requirements of the OIDCs have been put forward in recent years. These take into account the requirements mainly for financing developments in the energy sector and do not consider the immediate compulsion of foreign exchange requirements of oil imports. One such forecast is shown in Table 8.4, which estimates a total of 127.3 billion (in 1976 dollars) as being required by the OIDCs for investments in the energy sector. This staggering requirement would be a very high fraction of

Table 8.4. Projected non-OPEC LDC energy financing requirements, 1975–2000 (million barrels per day oil equivalent/billion 1976 US dollars).

	Oil: Production	Oil: Refining	Coal	Gas	Primary Electric	Total
2000 Capacity^a	14.4	14.4	7.8	2.1	1.5	
1975 Capacity^b	5.5	11.0	2.5	0.7	0.5	
	8.9	3.4	5.3	1.4	1.0	20.0 MMBDOE
Investment per capacity daily barrel	$5000	$1800	$5000	$3000	$46000	
Investment required (billion 1976$)	$44.5	$6.1	$26.5	$4.2	$46	$127.3 billion

Source: Brookhaven National Laboratory; Energy Needs, Uses and Resources in Developing Countries, 1978.

^{a}2000 capacity requirement equals projected year 2000 non-OPEC LDC consumption (total of all country groups).

^{b}1975 capacity is actual reported projection capacity. Difference (additional capacity required) times capital cost per daily barrel capacity (from preceding table) equals investment required to have sufficient projection capacity to satisfy projected year 2000 consumption.

the national incomes of the OIDCs as a whole and cannot possibly be met through domestic savings. For instance, currently in India's Sixth Five Year Plan, of the total outlay proposed, 30 per cent is planned to be invested in the energy sector, and even this will not remove some of the shortages that exist, particularly in the country's power sector. The response by multilateral organisations such as the World Bank has been heartening, but falls very short of the requirement, and some change in priorities is also highly desirable. Current lending by the World Bank for energy projects in the LDCs is only of the order of $1 billion annually, and even though the World Bank has now started funding oil projects which traditionally were open only to bilateral arrangements, its priorities and those of other funding organisations must very carefully match the resource endowments of beneficiary countries, or else the utilisation of energy resources may suffer from economic distortions and inefficiencies. There is for instance the important question of how much to invest in power development. One argument put forward is that a certain over-investment is desirable or else industrial activity may suffer and costly imports may have to be arranged. But the actual answer can be provided only on the basis of rigorous analysis. Since energy planning is a relatively new field, expertise developed in the industrial countries can suitably strengthen the process of planning in the OIDCs.

Whereas in the medium term development of conventional sources of energy acquires high priority, in the long run technologies will have to be developed and transferred for the utilisation of unconventional forms of energy. A study by the Solar Energy Research Institute discovered that approximately $225 million worth of projects dealing with renewable energy systems were currently being funded by developed countries. It was also estimated that grants and lending for renewable energy projects by the major OECD donor countries will reach $200 million within 5 years. This is a good beginning but may not be adequate unless effective institutions are developed to test new technologies and evaluate them objectively in the beneficiary countries themselves. Suitable institutional arrangements, therefore, have to be developed for testing and transferring new energy technologies over the next decade so that a significant impact is made by the turn of the century. In this regard, the OIDCs themselves can do a great deal through cooperation. Some form of cooperation would be desirable even for exploitation of conventional sources of energy. One such case would be existence of a large hydro-electric potential in Nepal, which can be tapped effectively only through a suitable regional arrangement.

CONCLUSIONS

Even though in tangible terms the energy prospects of the OIDCs do not appear promising, the only basis for cautious optimism is the growing awareness of the OIDC problem in the developed world. There is serious concern and a strong desire for action manifest in the deliberations of important international bodies.

The independent Commission on International Development Issues set up in September 1977 and headed by former West German Chancellor Willy Brandt has addressed and debated some of the problems discussed earlier and come up with a set of imaginative and sweeping recommendations with considerable emphasis on energy issues. A call has been made for the greater involvement of developing countries in international trade and finance toward an increased responsibility in the transfer of resources vital to future development. An Emergency Programme for 1980-85 aims to raise the contribution to development assistance by the wealthy industrial countries from an average of a little over 0.3 per cent to 0.7 per cent of their GNP. Further financial flows are thought possible through a projected system of international tax levies on trade, military expenditure and revenues from global resources such as minerals from the seabed. Provisions are also made for a more responsible approach by consumers and producers to the exploitation of oil to ensure the stability of supply, conservation, gradual and predictable price increases in real terms, and a more serious development of alternative energy sources.

If only two of the provisions of the Brandt Commission are implemented, then the situation for the OIDCs could improve substantially. An increase in development aid up to 0.7 per cent of GNP as proposed would mean an additional $40-50 billion annually, a substantial portion of which could be used for oil imports and indigenous energy development in the OIDCs. Further, if a tax of $1.00 per barrel is imposed on oil imported by the developed countries, this itself would lead to a total excess of $12.00 billion annually at current import levels, which would then be available for disbursement to the OIDCs. The Commission has also recommended that the level of 1 per cent of GNP as development assistance laid down by the Pearson report be achieved by the developed countries by 2000 AD.

How many of these recommendations are translated into action is yet to be seen, but the future of humanity lies in imaginative steps to ameliorate the problems of the less privileged. It has to be accepted by all that we are in this crisis together.

REFERENCES

1. Ashworth, John, Renewable Energy for the World's Poor, *Technology Review*, November 1979.
2. Barrie, T.W., and Lessile, E., Energy Policy in Developing Countries, *Energy Policy*, Vol. 6(2), pp. 119-128, June 1978.
3. Friedman, Efrain, Financing Energy in the Developing Countries, *Energy Policy*, Vol. 4(1), pp. 37-49, March 1976.
4. Grauer, Peter T., Capital Requirement of World's Oil Industry Over the Next Decade, *Petroleum Economist*, Vol. XLVI (7), p. 276, July 1979.
5. Energy in the Third World—Future Energy of India—a Delphi Study, *Energy Policy*, Vol. (4)(1) 1976.
6. Hoffman, R.T., The Re-appraisal of Multilateral Energy Aid, *Energy Policy*, Vol. (6)(4), pp. 332-338, December 1978.
7. Jenkins, Norman, 10th World Energy Conference Report (1977), *Energy Policy*, Vol. 5(4), December 1977.

8. Lambertini, Adrian, World Energy Prospects and the Developing World, *Finance & Development*, Vol. 16(4), pp. 18-22, December 1979.
9. Nathans, Robert & Palmedo, Philip, F., *Energy Planning and Management in Developing Countries*, Brookhaven National Laboratory, 11 October, 1977.
10. *Report of the Policy Analysis Division, Energy Needs, Uses and Resources in Developing Countries*, Brookhaven National Laboratory, March 1978.
11. *Report of the Working Group on Energy Policy*, Government of India, Controller of Publications, Delhi, 1980.

Chapter 9

Financing Energy Development in the Industrialised Countries

Paul Tempest*

This chapter reviews briefly various factors which will determine the capital requirements for energy of the industrialised countries over the next two decades, and tries to identify some of the hurdles to be overcome.

Current energy requirements for capital within the OECD countries are running at about $130-150 billion pa. This is equivalent to less than 2 per cent of the gross domestic product of the OECD area.

The task of securing the main technological advance to new forms of energy and to increased energy efficiency is likely to fall almost entirely to the leading industrialised countries who are best placed to deploy the necessary skills and who represent the bulk of demand (*see Table 9.1*).

Given the growing importance of energy in the western economy, despite its still low weight in industrial production or domestic consumption, it would seem at first sight hardly possible that such essential development could be constrained by an overall lack of capital. Nevertheless, every single energy project requires a mix of capital more or less appropriate to its own peculiar circumstances.

*Bank of England.

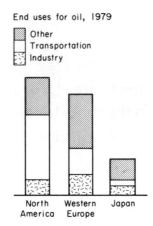

End uses for oil, 1979

▨ Other
☐ Transportation
░ Industry

North   Western   Japan
America Europe

Figure 9.1.

Let me begin therefore by examining the constraints and limits on the main sources of capital.

The four main sources of capital are:

(a) **Retentions within the energy sector.** The oil sector has traditionally relied on a very high level of profit retention and the ability of the leading multinational oil companies to transfer profits earned from one production area to a new one. The retained profit of the seven major oil multinationals was some $15 billion in 1979, ie only about 10 per cent of the total capital requirement of the OECD energy sector. Total net retentions of the energy sector also include those industries, nationalised and otherwise, which are authorised to plough back profit; the share of the nationalised industries in total retentions seems likely to increase.

(b) **Capital market sources.** Most energy borrowers use the international bond markets and internationally syndicated bank loans as well as domestic loan and bond issues, and the national equity markets. In 1979, international financial markets met total capital requirements of about $150 billion. Domestic financial markets probably accounted for something less than this figure. The share of energy in the total of $250–$300 billion was probably at least $25–$30 billion, ie over 10 per cent, although it is quite difficult to arrive at a more precise estimate from the statistics available, as many industrial borrowers for energy projects are classified under their corporate identity and not according to the

Table 9.1. World primary energy consumption (million tonnes oil equivalent)

	1969	*1974*	*1979*
United States	1,627	1,769	898
Canada	153	198	223
France	146	198	223
Italy	111	137	148
United Kingdom	208	215	221
West Germany	223	256	285
Japan	253	353	381
Total 7 above	2,721	3,112	3,351
U.S.S.R., E. Bloc and China	1,277	1,685	2,167
All other	903	1,174	1,442
Total world	4,901	5,971	6,960

Source: BP statistical review of the world oil industry 1979 (published July 1980).

purpose of their investment. In any case this share, which historically has been low, is now growing strongly.

(c) **International agency sources.** International agencies such as the World Bank and regional agencies such as the European Investment Bank are likely to play an increasing role within and outside the leading industrial countries. The bulk of their funds for lending, however, are likely to be raised on the world capital markets rather than from national subscription, and are likely to be directed towards the developing countries.

(d) **Government sources.** By far the largest source of capital for energy will remain that provided directly by Governments. Conservation investment, expenditure on research and development and new investment in high cost energy are all likely to depend increasingly on Government support.

The implications of different energy requirements

The OECD industrialised countries as a group depend on imports for one-third of their energy supply yet the dependence of individual countries varies greatly; for some, strategic considerations are as important as domestic or external energy costs; for others the structure of their industries and indigenous energy resources or

Table 9.2. Consumption of Primary Energy.

	Coal	Hydroelectricity	Nuclear	Natural Gas	Oil
North America	18%	6%	2%	29%	45%
Western Europe	19%	6%	2%	15%	58%
Japan	15%	6%	2%	3%	74%

Source: OECD World Energy Outlook and other publications (based on the statistics for 1976).

social/employment factors constitute powerful constraints. Competing national requirements for capital for energy development therefore constitute a real risk; co-ordination of energy investment policies along lines of mutual interest is clearly desirable, but difficult to achieve.

The risks of persistent low growth

Investment is almost always very closely related to patterns of economic growth. The fear of failing to break away from a low-growth pattern is itself a deterrent in the mobilisation of capital for the energy sector where most investment is front-end loaded and requires long lead-times. Compensatory action by governments in the wake of price discontinuities often merely exacerbates the problem, often placing sharp restraints on government spending and forcing up interest rates.

Price discontinuity as a throttle on investment

Abrupt changes in the perceived commercial viability of alternative energy and high-cost non-OPEC conventional energy have become a feature of the 1970s. Each jump in the oil price appears to give a rationale for heavy investment in alternatives, yet the intervening declining real price seriously erodes confidence. There is widespread fear that this cycle of developments could go on repeating itself. Much of the international economic, and financial debate over the last decade has therefore been taken up with the issue of purging the economic system of these discontinuities, whether in the short term by strategic stockpiling or in the long term by negotiated settlement between the producers and consumers or by bringing other economic and political forces to bear.

Table 9.3. Two estimates of energy production costs (U.S. dollars per barrel oil equivalent)

	Workshop on Alternative Energy Sources, 1977	Shell, 1980
	1976 U.S. dollars	*Mid-1980 U.S. dollars*
Oil		
Arabian light	0.25	Conventional oil (Middle East) 1–3
North Sea	5–8	5–15
Electricity		
Conventional, thermal and nuclear	40–70	60–130
Wind (on site 1985)	50+	(Possibly on site 1980) 120+
Photovoltaic solar (on site 1985)	120+	(Possibly on site 1980s) 250+
Thermal Energy Alternatives		
Imported coal (EEC)	6–10	10–15
Indigenous coal (EEC)	8–15	10–20
Nuclear breakeven	5–10	10–20
LNG imports	10–20	25–35
Liquids from shale/(U.S.) tar sands	15–25	15–30
Liquids from imported coal (EEC)	25–35	46–65
Biomass (crops for fuel)	40–50	45–80+

The investment climate

Several themes emerge from this brief review. The capital requirements of the industrial countries for energy investment over the next two decades are likely to be much greater in both absolute and relative terms. than hitherto. Nonetheless, the western industrialized countries are amply equipped to meet the task of mobilising the necessary technological and financial resources, provided the investment climate is stable. No obstacle to investment can be foreseen from outright lack of capital, either from the official or private sources. The potential therefore exists to effect a massive substitution of capital for energy through conservation, and increased efficiency of energy use, and, through a further deployment of capital, to achieve optimum inter-fuel substitution and to develop new alternative sources of energy. In the long run, energy available at a stable and predictable price, may once again be an important factor in economic growth.

REFERENCES

1. Tempest, Paul. Capital Requirements for Energy in the Industrialised Countries. *Revue de L'Énergie*, Special Issue for the World Energy Conference: Energy Investment Finance, September 1980.
2. Berndt, E.R., and Wood, W. Technology, Prices and the Derived Demand for Energy, *Review of Economics and Statistics*, 57, August 1975.
3. *BP Statistical Review of the World Oil Industry*, 1979.
4. Dunkerley, J. (Editor). International Energy Strategies. Proceedings of the 1979 Annual International Conference of the IAEE, Washington D.C. Oelgeschlager, Gunn and Hain, 1980.
5. Exxon, *World Energy Outlook*, December 1979.
6. Houthakker, H.S. New Evidence on Demand Elasticities, *Econometrica* 33, pp. 277-288.
7. Jorgensen, D.E. Consumer Demand for Energy. In *International Studies of the Demand for Energy*. Edited by W.D. Nordhaus, North Holland Publishing Company, Amsterdam, 1977.
8. Odell, P. R. World Energy in the 1980s. *Scottish Journal of Political Economy*, November 1979.
9. Organisation for Economic Co-operation and Development, Energy Prospects to 1985, Paris 1974. *World Energy Outlook*, 1977.
10. Oystein, N.O. Oil Politics in the 1980s, Council on Foreign Relations, 1979.
11. Pindyck, R. S. *The Structure of World Energy Demand*, MIT Press, 1979.
12. Shell International Petroleum Company. *Energy and the Investment Challenge*, September 1979.
13. U.K. Department of Energy, Report of the Working Group on Energy Elasticities, *Energy Paper No. 17*, February 1977.
14. Workshop on Alternative Energy Strategies, *Energy Global Prospects 1985-2000*. McGraw Hill, New York 1977.

PART V

New Approaches to Productivity and Employment

Chapter 10

Energy Prices and Productivity Growth

Dale W. Jorgenson *

INTRODUCTION

The growth of the U.S. economy in the postwar period has been very rapid by historical standards. The rate of economic growth reached its maximum during the period 1960 to 1966. Growth rates have slowed substantially since 1966 and declined further since 1973. A major source of uncertainty in projections of the future of the U.S. economy is whether patterns of growth will better conform to the rapid growth of the early 1960s, the more moderate growth of the late 1960s and early 1970s or the disappointing growth since 1973.

In this paper our first objective is to identify the sources of uncertainty about future U.S. economic growth more precisely. For this purpose we decompose the growth of output during the postwar period into contributions of capital input, labor input, and productivity growth. For the period 1948 to 1976 we find that all three sources of economic growth are significant and must be considered in analyzing future growth potential. For the postwar period capital input has made the most important contribution to the growth of output, productivity growth has been next most important, and labor input has been least important.

*Professor, Harvard University.

Focusing on the period 1973 to 1976, we find that the fall in the rate of economic growth has been due to a dramatic decline in productivity growth. Declines in the contributions of capital and labor input are much less significant in explaining the slowdown. We conclude that the future growth of productivity is the main source of uncertainty in projections of future U.S. economic growth.

Our second objective is to analyze the slowdown in productivity growth for the U.S. economy as a whole in greater detail. For this purpose we decompose productivity growth during the postwar period into components that can be identified with productivity growth at the sectoral level and with reallocations of output, capital input, and labor input among sectors. For the period 1948 to 1976, we find that these reallocations are insignificant relative to sectoral productivity growth. The combined effect of all three reallocations is slightly negative, but sufficiently small in magnitude to be negligible as a source of aggregate productivity growth.

Again focusing on the period 1973 to 1976, it is possible that the economic dislocations that accompanied the severe economic contraction of 1974 and 1975 could have resulted in shifts of output and inputs among sectors that contributed to the slowdown of productivity growth at the aggregate level. Alternatively, the sources of the slowdown might be found in slowing productivity growth at the level of individual industrial sectors. We find that the contribution of reallocations of output and inputs among sectors was positive rather than negative during the period 1973–1976 and relatively small. Declines in productivity growth for the individual industrial sectors of the U.S. economy must bear the full burden of explaining the slowdown in productivity growth for the economy as a whole.

The decomposition of the growth of output among contributions of capital input, labor input, and productivity growth is helpful in isolating the sources of uncertainty in future growth projections. The further decomposition of productivity growth among the reallocations of output, capital input, and labor input among sectors and growth in productivity at the sectoral level provides additional detail. The uncertainty in future growth projections can be resolved only by providing an explanation for the fall in productivity growth at the sectoral level. For this purpose an econometric model of sectoral productivity growth is required.

Our third objective is to present the results of an econometric analysis of the determinants of productivity growth at the sectoral level. Our econometric model determines the growth of sectoral productivity as a function of relative prices of sectoral inputs. For each sector we divide inputs among capital, labor, energy, and materials inputs. We allow for the fact that the value of sectoral output includes the value of intermediate inputs—energy and materials—as well as the value of primary factors of production—capital and labor. Differences in relative prices for inputs are associated with differences in productivity growth for each sector.

After fitting our econometric model of productivity growth to data for individual industrial sectors we find that productivity growth decreases with an increase in the price of capital input for a very large proportion of U.S. industries. Similarly, productivity growth falls with higher prices of labor input for a large proportion of industries. The impact of higher energy prices is also to slow the growth of productivity for a large proportion of industries. By contrast we find that an increase in the price of materials input is associated with increases in productivity growth for almost all industries.

Since 1973 the relative prices of capital, labor, energy, and materials inputs have been altered radically as a consequence of the increase in the price of energy relative to other productive inputs. Higher world petroleum prices following the Arab oil embargo of late 1973 and 1974 have resulted in sharp increases in prices for all forms of energy in the U.S. economy—oil, natural gas, coal, and electricity generated from fossil fuels and other sources. Although the U.S. economy has been partly shielded from the impact of higher world petroleum prices through a system of price controls, all industrial sectors have experienced large increases in the price of energy relative to other inputs.

Our econometric model reveals that slower productivity growth at the sectoral level is associated with higher prices of energy relative to other inputs. Our first conclusion is that the slowdown of sectoral productivity growth after 1973 is a consequence of the sharp increase in the price of energy relative to other productive inputs that began with the run-up of world petroleum prices in late 1973 and early 1974. The fall in sectoral productivity growth after 1973 is responsible in turn for the decline in productivity growth for the U.S. economy as a whole. Slower productivity growth is the primary source of the slow-down in U.S. economic growth since 1971.

Our final objective is to consider the prospects for future U.S. economic growth. Exports of petroleum from Iran dropped sharply during 1979, following the revolution in that country in late 1978. During 1979 world petroleum prices have jumped 130 to 140 percent, resulting in large and rapid price increases for petroleum products in the United States. During 1979 the prices of petroleum products began to move to world levels as a consequence of the decontrol of domestic prices by the U.S. government over the period 1979 to 1981. Prices of natural gas will also be allowed to rise through decontrol by 1985 or, at the latest, by 1987. Prices of energy confronted by individual industries within the United States have already increased relative to other productive inputs and can be expected to increase further.

Based on the performance of the U.S. economy since 1973, we can anticipate a further slowdown in the rate of economic growth, a decline in the growth of productivity for the economy as a whole, and declines in sectoral productivity growth for a wide range of industries. These dismal conclusions suggest that a return to rapid growth of the early 1960s is highly unlikely, that even the slower growth of the late 1960s and early 1970s will be difficult to attain, and that the performance of the U.S. economy during the 1980s could be worse than during the period from 1973 to the present. We conclude the paper with a discussion of policy measures to ameliorate the negative effects of higher energy prices on future U.S. economic growth.

THE GROWTH SLOWDOWN

In this section we begin our analysis of the slowdown in U.S. economic growth by decomposing the growth of output for the economy as a whole into the contributions of capital input, labor input, and productivity growth.[1] The results are given in Table 10.1 for the postwar period 1948-1976 and for the following seven subperiods—1948-1953, 1953-1957, 1957-1960, 1960-1966, 1966-1969, 1969-1973, and 1973-1976.[2] Except for the period from 1973 to 1976, each of the subperiods covers economic activity from one cyclical peak to the next. The last period covers economic activity from the cyclical peak in 1973 to 1976, a year of recovery from the sharp downturn in economic activity in 1974 and 1975.

We first present rates of growth for output, capital input, labor input, and productivity for the U.S. economy. For the postwar period as a whole output grew at 3.50 percent per year, capital input grew at 4.01 percent, and labor grew at 1.28 percent. The growth of productivity averaged 1.14 percent per year. The rate of economic growth reached its maximum at 4.83 percent during the period 1960-1966 and grew at only 0.89 percent during the recession and partial recovery of 1973-1976. The growth of capital output was more even, exceeding 5 percent in 1948-1953 and 1966-1969 and falling to 3.12 percent in 1973-1976. The growth of labor input

Table 10.1, Growth of output and inputs for the U.S. economy, 1948-1976

	1948- 1976	*1948- 1953*	*1953- 1957*	*1957- 1960*	*1960- 1966*	*1966- 1969*	*1969- 1973*	*1973- 1976*
Growth rates:								
Output	0.0350	0.0457	0.0313	0.0279	0.0483	0.0324	0.0324	0.0089
Capital input	0.0401	0.0507	0.0393	0.0274	0.0376	0.0506	0.0396	0.0312
Labor input	0.0128	0.0160	0.0023	0.0099	0.0199	0.0185	0.0116	0.0058
Productivity	0.0114	0.0166	0.0146	0.0113	0.0211	0.0004	0.0095	-0.0070
Contributions:								
Capital input	0.0161	0.0194	0.0154	0.0109	0.0156	0.0211	0.0161	0.0126
Labor input	0.0075	0.0097	0.0013	0.0057	0.0116	0.0108	0.0068	0.0033

reached its maximum in the period 1960-1966 at 1.99 percent and fell to 0.58 percent in 1973-1976, which was above the minimum of 0.23 percent in the period 1953-1957.

We can express the rate of growth output for the U.S. economy as a whole as the sum of a weighted average of the rates of growth of capital and labor inputs and the growth of productivity. The weights associated with capital and labor inputs are average shares of these inputs in the value of output. The contribution of each input is the product of the average share of this input and correspond-ing in ⁚t growth rate. We present contributions of capital and labor inputs ʋo U.S. economic growth for the period 1948-1976 and for seven subperiods in Table 10.1. Considering productivity growth, we find that the maximum occurred from 1960 to 1966 at 2.11 percent per year. During the period 1966-1969 productivity growth was almost negligible at 0.04 percent. Productivity growth recovered to 0.95 percent during the period 1969-1973 and fell to a negative 0.70 percent during 1973-1976.

Since the value shares of capital and labor inputs are very stable over the period 1948-1976, the movements of the contributions of these inputs to the growth of output largely parallel those of the growth rates of the inputs themselves. For the postwar period as a whole the contribution of capital input of 1.61 percent is the most important source of output growth. Productivity growth is next most important at 1.14 percent, while the contribution of labor input is the third most important at 0.75 percent. All three sources of growth are significant and must be considered in an analysis of the slowdown of economic growth during the period 1973-1976. However, capital input is clearly the most important contributor to the rapid growth of the U.S. economy during the postwar period.[3]

Focusing on the period 1973 to 1976, we find that the contribution of capital input fell to 1.26 percent for a drop of 0.35 percent from the postwar average, the contribution of labor input fell to 0.33 percent for a drop of 0.42 percent, and that productivity growth at a negative 0.70 percent dropped 1.84 percent. We conclude that the fall in the rate of U.S. economic growth during the period 1973–1976 was largely due to the fall in productivity growth. Declines in the contributions of the capital and labor inputs are much less significant in explaining the slowdown. A detailed explanation of the fall in productivity growth is needed to account for the slowdown in U.S. economic growth.

To analyze the sharp decline in productivity growth for the U.S. economy as a whole during the period 1973 to 1976 in greater detail we employ data on productivity growth for individual industrial sectors. For this purpose it is important to distinguish between productivity growth at the aggregate level and productivity growth at the sectoral level. At the aggregate level the appropriate concept of output is value added, defined as the sum of the values of capital and labor inputs for all sectors of the economy. At the sectoral level the appropriate concept of output includes the value of primary factors of production at the sectoral level—capital and labor inputs—and the value of intermediate inputs—energy and materials inputs. In aggregating over sectors to obtain output for the U.S. economy as a whole the production and consumption of intermediate goods cancel out, so that values of energy and materials inputs do not appear at the aggregate level.

We can express productivity growth for the U.S. economy as a whole as the sum of four components. The first component is a weighted sum of productivity growth rates for individual industrial sectors. The weights are ratios of the value of output in each sector to value added in that sector. The sum of these weights over all sectors exceeds unity, since productivity growth in each sector contributes to the growth of output in that sector and to the growth of output in other sectors through deliveries of intermediate inputs to those sectors. The remaining components of aggregate productivity growth represent the contributions of reallocations of value added, capital input, and labor input among sectors to productivity growth for the economy as a whole.[4]

The role of reallocations of output, capital input and labor input among sectors is easily understood. For example, if capital input moves from a sector with a relatively low rate of return to a sector with a high rate of return, the quantity of capital input for the economy as a whole is unchanged, but the level of output is increased, so that productivity has improved. Similarly, if labor input moves from a sector with low wages to a sector with high wages, labor input is unchanged, but productivity has improved. Productivity growth for the economy as a whole is a combination of improvements in productivity at the sectoral and reallocations of output, capital input and labor input among sectors. Data on reallocations of output, capital input, and labor input for the postwar period 1948 to 1976 and for seven subperiods are given in Table 10.2.[5]

Table 10.2. Productivity growth for the U.S. economy 1948-1976.

	1948– 1976	1948– 1953	1953– 1957	1957– 1960	1960– 1966	1966– 1969	1969– 1973	1973– 1976
Sectoral productivity growth	0.0124	0.0219	0.0177	0.0145	0.0217	0.0025	0.0048	0.0113
Reallocation of value added	–0.0016	–0.0076	–0.0030	–0.0010	–0.0016	–0.0025	0.0030	0.0046
Reallocation of capital input	0.0008	0.0022	0.0008	–0.0001	0.0002	0.0001	0.0010	0.0008
Reallocation of labor input	–0.0002	–0.000	–0.0008	–0.0021	0.0008	0.0004	0.0006	–0.0011

For the postwar period as a whole productivity growth at the aggregate level is dominated by the contribution of sectoral productivity growth of 1.24 percent per year. The contributions of reallocations of output, capital input, and labor input are a negative 0.16 percent, a positive 0.08 percent, and a negative 0.02 percent. Adding these contributions together we find that the combined effect of the three reallocations is a negative 0.10 percent, which is negligible by comparison with the effect of productivity growth at the sectoral level. Productivity growth at the aggregate level provides an accurate picture of average productivity growth for individual industries; this picture is not distorted in an important way by the effect of reallocations of output and inputs among sectors.

Again focusing on the period 1973-1976, we find that the contribution of sectoral productivity growth to productivity growth for the economy as a whole fell to a negative 1.13 percent for a drop of 2.37 percent from the postwar average. By contrast the contribution of reallocations of output rose to 0.46 percent for a gain of 0.62 percent from the postwar average. The contribution of the reallocation of capital input was unchanged at 0.08 percent, while the contribution of labor input fell to a negative 0.11 percent for a drop of 0.09 percent from the postwar average. The combined contribution of all three reallocations rose 0.53 percent, partially offsetting the precipitous decline in productivity growth at the sectoral level. We conclude that declines in productivity growth for the individual industrial sectors of the U.S. economy are more than sufficient to explain the decline in productivity growth for the economy as a whole.

To summarize our findings on the slowdown of U.S. economic growth during the period 1973–1976, we find that the drop in the growth of output of 2.61 percent per year from the postwar average is the sum of a decline in the contribution of labor input of 0.42 percent per year, a sharp dip in sectoral rates of productivity growth of 2.37 percent, a rise in the role of reallocations of output among sectors of 0.62 percent per year, no change in the reallocations of capital input, and a decline in the contribution of reallocations of labor input of 0.09 percent per year. Whatever the causes of the slowdown, they are to be found in the collapse of productivity growth at the sectoral level rather than a slowdown in the growth of capital and labour inputs at the aggregate level or the reallocations of output, capital input, or labor input among sectors.

The decomposition of economic growth into the contributions of capital input, labor input, and productivity growth is helpful in pinpointing the causes of the slowdown. The further decomposition of productivity growth for the economy as a whole into contributions of sectoral productivity growth and reallocations of output, capital input, and labor input is useful in providing additional detail. However, our measure of sectoral productivity growth is simply the unexplained residual between growth of sectoral output and the contributions of sectoral capital, labor, energy, and materials inputs. The problem remains of providing an explanation for the fall in productivity growth at the sectoral level.

SECTORAL PRODUCTIVITY GROWTH

We have now succeeded in identifying the decline in productivity growth at the level of individual industrial sectors within the U.S. economy as the main culprit in the slowdown of U.S. economic growth that took place after 1973. To provide an explanation for the slowdown we must go behind the measurements to identify the determinants of productivity growth at the sectoral level. For this purpose we require an econometric model of sectoral productivity growth. In this section we present a summary of the results of applying such an econometric model to detailed data on sectoral output and capital, labor, energy, and materials inputs for thirty-five individual industries in the United States.

Our complete econometric model is based on sectoral price functions for each of the thirty-five industries included in our study.[6] Each price function gives the price of the output of the corresponding industrial sector as a function of the prices of capital, labor, energy, and materials inputs and time, where time represents the level of technology in the sector.[7] Obviously, an increase in the price of one of the inputs, holding the prices of the other inputs and the level of technology constant, will necessitate an increase in the price of output. Similarly, if productivity in a sector improves and the prices of all inputs into the sector remain the same, the price of output must fall. Price functions summarize these and other relationships among the prices of output, capital, labor, energy, and materials inputs, and the level of technology.

Although the sectoral price functions provide a complete model of production patterns for each sector, it is useful to express this model in an alternative and equivalent form. We can express the shares of each of the four inputs—capital, labor, energy, and materials —in the value of output as functions of the prices of these inputs and time, again representing the level of technology.[8] We can add to these four equations for the value shares an equation that expresses productivity growth as a function of the prices of the four inputs and time.[9] In fact, the negative of the rate of productivity growth is a function of the four input prices and time. This equation is our econometric model of sectoral productivity growth.[10]

Like any econometric model, the relationships determining the value shares of capital, labor, energy, and materials inputs and the negative of the rate of productivity growth involve unknown parameters that must be estimated from data for the individual industries. Included among these unknown parameters are biases of productivity growth that indicate the effect of changes in the level of technology on the value shares of each of the four inputs.[11] For example, the bias of productivity growth for capital input gives the change in the share of capital input in the value of output in response to changes in the level of technology, represented by time. Similarly, biases of productivity growth for labor, energy, and materials inputs give changes in the shares of labor, energy, and materials inputs in the value of output that results from changes in the level of technology.

We say that productivity growth is capital using if the bias of productivity growth for capital input is positive, that is, if changes in the level of technology result in an increase in the share of capital input in the value of output, holding all input prices constant. Productivity growth involves an increase in the quantity of capital input as technology changes, so that we say that the change in technology is capital using. Similarly, we say that productivity growth is capital saving if the bias of productivity growth for capital input is negative. As technology changes, the production process uses less capital input, so that the change in technology is capital saving.

Similarly, we can say that productivity growth is labor using or labor saving if the bias of productivity growth for labor input is positive or negative. As technology changes, the production process uses more or less labor input, depending on whether the change in technology is labor using or labor saving. We can associate energy using or energy saving productivity growth with positive or negative biases or productivity growth for energy input. Finally, we can associate materials using or materials saving productivity growth with positive or negative biases of productivity growth for materials input. Since the shares of all four inputs—capital, labor, energy, and materials—sum to unity, productivity growth that "uses" or "saves" all four inputs is impossible. In fact, the sum of the biases for all four must be precisely zero, since the changes in all four shares with any change in technology must sum to zero.

We have pointed out that our econometric model for each industrial sector of the U.S. economy includes an equation giving the negative of sectoral productivity growth as a function of the prices of the four inputs and time. The biases of technical change with respect to each of the four inputs appear as the coefficients of time, representing the level of technology, in the four equations for the value shares of all four inputs. The biases also appear as coefficients of the prices in the equation for the negative of sectoral productivity growth. This feature of our econometric model makes it possible to use information about changes in the value shares with time and changes in the rate of sectoral productivity growth with prices in determining estimates of the biases of technical change.

The biases of productivity growth express the dependence of value shares of the four inputs on the level of technology and also express the dependence of the negative of productivity growth on the input prices. We can say that capital using productivity growth, associated with a positive bias of productivity growth for capital input, implies that an increase in the price of capital input decreases the rate of productivity growth (or increases the negative of the rate of productivity growth). Similarly, capital saving productivity growth, associated with a negative bias for capital input, implies that an increase in the price of capital input increases the rate of productivity growth. Analogous relationships hold between biases of labor, energy, and materials inputs and the direction of the impact of changes in the prices of each of these inputs on the rate of productivity growth.[12]

Jorgenson and Fraumeni [1980] have fitted biases of productivity growth for thirty-five industrial sectors that make up the whole of the producing sector of the U.S. economy. They have also fitted the other parameters of the econometric model that we have described above. Since our primary concern in this section is to analyze the determinants of productivity growth at the sectoral level, we focus on the patterns of productivity growth revealed in Table 10.3. We have listed the industries characterized by each of the possible combinations of biases of productivity growth, consisting of one or more positive biases and one or more negative biases.[13]

The pattern of productivity growth that occurs most frequently in Table 10.3 is capital using, labor using, energy using, and materials saving productivity growth. This pattern occurs for nineteen of the thirty-five industries analyzed by Jorgenson and Fraumeni. For this pattern of productivity growth the bias of productivity growth for capital input, labor input, and energy input are positive, and the bias of productivity growth for materials input is negative. This pattern implies that increases in the prices of capital input, labor input, and energy input decrease the rate of productivity growth, while increases in the price of materials input increase the rate of productivity growth.

Considering all patterns of productivity growth included in Table 10.3, we find that productivity growth is capital using for twenty-five of the thirty-five industries included in our study. Productivity growth is capital saving for the remaining ten industries. Similarly, productivity growth is labor using for thirty-one of the thirty-five industries and labor saving for the remaining four industries; productivity growth is energy using for twenty-nine of the thirty-five industries included in Table 10.3 and is energy saving for the remaining six. Finally, productivity growth is materials using for only two of the thirty-five industries and is materials saving for the remaining thirty-three. We conclude that for a very large proportion of industries the rate of productivity growth decreases with increases in the prices of capital, labor, and energy inputs, and increases in the price of materials inputs.

The most striking change in the relative prices of capital, labor, energy and materials inputs that has taken place since 1973 is the

Table 10.3. Classification of industries by biases of productivity growth.

Pattern of biases	Industries
Capital using *labor using* *energy using* *material saving*	agriculture, metal mining, crude petroleum and natural gas, non-metallic mining, textiles, apparel, lumber, furniture, printing, leather, fabricated metals, electrical machinery, motor vehicles, instruments, miscellaneous manufacturing, transportation, trade, finance, insurance and real estate, services.
Capital using *labor using* *energy saving* *material saving*	Coal mining, tobacco manufactures, communications, government enterprises.
Capital using *labor saving* *energy using* *material saving*	Petroleum refining.
Capital using *labor saving* *energy saving* *material using*	Construction.
Capital saving *labor saving* *energy using* *material saving*	Electric utilities.
Capital saving *labor using* *energy saving* *material saving*	Primary metals.
Capital saving *labor using* *energy using* *material saving*	Paper, chemicals, rubber, stone, clay and glass, machinery except electrical, transportation equipment and ordnance, gas utilities.
Capital saving *labor saving* *energy using* *material using*	Food.

staggering increase in the price of energy. The rise in energy prices began in 1972 before the Arab oil embargo, as the U.S. economy moved toward the double-digit inflation that characterized 1973. In late 1973 and early 1974 the price of petroleum on world markets increased by a factor of four, precipitating a rise in domestic prices of petroleum products, natural gas, coal, and uranium. The impact of higher world petroleum prices was partly deflected by price controls for petroleum and natural gas that resulted in the emergence of shortages of these products during 1974. All industrial sectors of the

U.S. economy experienced sharp increases in the price of energy relative to other inputs.

Slower growth in productivity at the sectoral level is associated with higher energy prices for twenty-nine of the thirty-five industries that make up the producing sector of the U.S. economy. The dramatic increases in energy prices resulted in a slowdown in productivity growth at the sectoral level. In the preceding section we have seen that the fall in sectoral productivity growth after 1973 is the primary explanation for the decline in productivity for the U.S. economy as a whole. Finally, we have shown that the slowdown in productivity growth during the period 1973–1976 is the main source of the fall in the rate of U.S. economic growth since 1973.

We have now provided a solution to the problem posed by the disappointing growth record of the U.S. economy since 1973. By reversing historical trends toward lower prices of energy in the U.S. economy, the aftermath of the Arab Oil Embargo of 1973 and 1974 has led to an end to rapid economic growth. The remaining task is to draw the implications of our findings for future U.S. economic growth. Projections of future economic growth must take into account the dismal performance of the U.S. economy since 1973 as well as the rapid growth that has characterized the U.S. economy during the postwar period. In particular, such projections must take into account the change in the price of energy input for individual industrial sectors, relative to prices of capital, labor, and materials inputs.

PROGNOSIS

Our objective in this concluding section of the paper is to provide a prognosis for future U.S. economic growth. For this purpose we cannot rely on the extrapolation of past trends in productivity growth or its components. The year 1973 marks a sharp break in trend associated with a decline in rates of productivity growth at the sectoral level. Comparing the period after 1973 with the rest of the postwar period, we can associate the decline in productivity growth with the dramatic increase in energy prices that followed the Arab oil embargo in late 1973 and early 1974. The remaining task is to analyze the prospects for a return to the high sectoral productivity growth rates of the early 1960s, for moderate growth of sectoral productivity growth like that of the late 1960s and early 1970s, or for continuation of the disappointing growth since 1973.

During 1979 there has been a further sharp increase in world petroleum prices, following the interruption of Iranian petroleum exports that accompanied the revolution that took place in that country in late 1978. Although prices of petroleum sold by different petroleum exporting countries differ widely, the average price of petroleum imported into the United States has risen by 130 to 140 percent since December 1978. In April 1979 President Carter announced that prices of petroleum products would be gradually decontrolled over the period from May 1979 to September 1981. As a consequence domestic petroleum prices in the United States will move to world levels in a relatively short period of time. Domestic natural gas prices will also be subject to gradual decontrol, moving to world levels as early as 1985 or, at the latest, 1987.

Given the sharp increase in the price of energy relative to the prices of other productive inputs, the prospects for productivity growth at the sectoral level are dismal. In the absence of any reduction in prices of capital and labor inputs during the 1980s, we can expect a decline in productivity growth for a wide range of U.S. industries, a decline in the growth of productivity for the U.S. economy as a whole, and a further slowdown in the rate of U.S. economic growth. To avoid a repetition of the unsatisfactory economic performances of the 1970s it is essential to undertake measures to reduce the price of capital input and labor inputs. The price of capital input can be reduced by cutting taxes on income from capital.[14] Similarly, payroll taxes can be cut in order to reduce the price of labor input.

The prospects for changes in tax policy that would have a substantial positive impact on productivity growth in the early 1980s are not bright. Any attempt to balance the Federal budget during 1981 in the face of a sharp recession during the last half of 1980 and the first half of 1981 will require tax increases rather than tax cuts. Higher inflation rates have resulted in an increase in the effective rate of taxation of capital. Payroll taxes are currently scheduled to rise in 1981. For these reasons it appears that a return to the rapid growth of the 1960s is out of the question. Even the moderate growth of the 1960s and early 1970s would be difficult to attain. In the absence of measures to cut taxes on capital and labor inputs, the performance of the U.S. economy during the 1980s could be worse than during the period from 1973 to the present.

For economists the role of productivity in economic growth presents a problem comparable in scientific interest and social importance to the problem of unemployment during the Great Depression of the 1930s. Conventional methods of economic analysis have been tried and have been found to be inadequate. Clearly, a new framework will be required for economic understanding. The findings we have outlined above contain some of the elements that will be required for the new framework for economic analysis as the U.S. economy enters the 1980s.

At first blush the finding that higher energy prices are an important determinant of the slowdown in U.S. economic growth seems paradoxical. In aggregative studies of sources of economic growth energy does not appear as an input, since energy is an intermediate good and flows of intermediate goods appear as both outputs and inputs of individual industrial sectors, canceling out for the economy as a whole.[15] It is necessary to disaggregate the sources to economic growth into components that can be identified with output and inputs at the sectoral level in order to define an appropriate role for energy.[16]

Within a framework for analyzing economic growth that is disaggregated to the sectoral level it is not sufficient to provide a decomposition of the growth of sectoral output among the contributions of sectoral inputs and the growth of sectoral productivity.[17] It is necessary to explain the growth of sectoral productivity by means of an econometric model of productivity growth for each sector. Without such econometric models the growth of sectoral productivity is simply an unexplained residual between the growth of output and the contributions of capital, labor, energy, and materials inputs.

Finally, the parameters of an econometric model of production must be estimated from empirical data in order to determine the direction and significance of the influence of energy prices on productivity growth at the sectoral level.[18] From a conceptual point of view a model of production is consistent with positive, negative, or zero impacts of energy prices on sectoral productivity growth. From an empirical point of view the influence of higher energy prices is negative and highly significant. There is no way to substantiate this empirical finding without estimates of the unknown parameters of the econometric model of productivity growth.

The steps we have outlined—disaggregating the sources of economic growth down to the sectoral level, decomposing the rate of growth of sectoral output into sectoral productivity growth and the contributions of capital, labor, energy, and materials inputs, and modeling the rate of growth of productivity econometrically—have been taken only recently. Much additional research will be required to provide an exhaustive explanation of the slowdown of U.S. economic growth within the new framework and to derive the implications of the slowdown for future growth of the economy.

NOTES

1. The methodology that underlies our decomposition of the growth of output is presented in detail by Jorgenson [1979].
2. The results presented in Table 10.1 are those of Fraumeni and Jorgenson [1979], who also provide annual data for output and inputs.
3. This conclusion contrasts sharply with that of Denison [1979]. For a comparison of our methodology with that of Denison, see Jorgenson and Griliches [1972].
4. The methodology that underlies our decomposition of productivity growth is presented in detail by Jorgenson [1979].
5. The results presented in Table 10.2 are those of Fraumeni and Jorgenson [1979], who also provide annual data for productivity growth.
6. Econometric models for each of the thirty-five industries are given by Jorgenson and Fraumeni [1980].
7. The price function was introduced by Samuelson [1953]. A complete characterization of the sectoral price functions employed in this study is provided by Jorgenson and Fraumeni [1980].
8. Our sectoral price functions are based on the translog price function introduced by Christensen, Jorgenson, and Lau [1971, 1973]. The translog price function was first applied at the sectoral level by Berndt and Jorgenson [1973] and Berndt and Wood [1975]. References to sectoral production studies incorporating energy and materials inputs are given by Berndt and Wood [1979].
9. Productivity growth is represented by the translog index introduced by Christensen and Jorgenson [1970]. The translog index of productivity growth was first derived from the translog price function by Diewert [1980] and by Jorgenson and Lau [1980].
10. This model of sectoral productivity growth is based on that of Jorgenson and Lau [1980].
11. The bias of productivity growth was introduced by Hicks [1932]. An alternative definition of the bias of productivity growth was introduced by Binswanger [1974a, 1974b]. The definition of the bias of productivity growth employed in our econometric model is due to Jorgenson and Lau [1980].
12. A complete characterization of biases of productivity growth is given by Jorgenson and Fraumeni [1980].
13. The results presented in Table 10.3 are those of Jorgenson and Fraumeni [1980]. Of the fourteen logically possible combinations of biases of productivity growth, only the eight patterns presented in Table 10.3 occur empirically.
14. An analysis of alternative proposals for cutting taxes on income from capital is presented by Auerbach and Jorgenson [1980].
15. See, for example, Denison [1979].
16. Kendrick [1961, 1973] has presented an analysis of productivity growth at the sectoral level. However, his measure of productivity growth is based on value added at the sectoral level, so that no role is provided for energy and materials inputs in productivity growth. For a more detailed discussion, see Jorgenson [1979].
17. Gollop and Jorgenson [1980] have presented an analysis of productivity growth at the sectoral level based on the concept of output that includes

both primary factors of production and intermediate inputs.
18. Estimates of the parameters of an econometric model of sectoral productivity growth are presented by Jorgenson and Fraumeni [1980].

REFERENCES

Auerbach, Alan J., and Dale W. Jorgenson [1980], "The First Year Capital Recovery System," *Harvard Business Review*, forthcoming.

Berndt, Ernst R., and Dale W. Jorgenson [1973], "Production Structure," Chapter 3 in Dale W. Jorgenson and Hendrik S. Houthakker, eds., *U.S. Energy Resources and Economic Growth*, Washington, Energy Policy Project.

Berndt, Ernst R., and David O. Wood [1975], "Technology, Prices, and the Derived Demand for Energy," *Review of Economics and Statistics*, Vol. 56, No. 3, August, pp. 259-268.

—— [1979], "Engineering and Econometric Interpretations of Energy-Capital Complementarity," *American Economic Review* Vol. 69, No. 3, September, pp. 342-354.

Binswanger, Hans P. [1974a], "The Measurement of Technical Change Biases With Many Factors of Production," *American Economic Review*, Vol. 64, No. 5, December, pp. 964-976.

—— [1974b], "A Microeconomic Approach to Induced Innovation," *Economic Journal*, Vol. 84, No. 336, December, pp. 940-958.

Christensen, Laurits R., and Dale W. Jorgenson [1970], "U.S. Real Product and Real Factor Input, 1929-1967," *Review of Income and Wealth*, Series 16, No. 1, March, pp. 19-50.

Christensen, Laurits R., Dale W. Jorgenson and Lawrence J. Lau [1971], "Conjugate Duality and the Transcendental Logarithmic Production Function," *Econometrica*, Vol. 39, No. 4, July, pp. 255-256.

—— [1973], "Transcendental Logarithmic Production Frontiers," *Review of Economics and Statistics*, Vol. 55, No. 1, February, pp. 28-45.

Denison, Edward F. [1979], *Accounting for Slower Economic Growth*. Washington: The Brookings Institution.

Diewert, W. Erwin [1980], "Aggregation Problems in the Measurement of Capital," in Dan Usher, ed., *The Measurement of Capital*, Chicago, University of Chicago Press, forthcoming.

Fraumeni, Barbara M. and Dale W. Jorgenson [1979], "The Role of Capital in U.S. Economic Growth, 1948-1976." in George M. von Furstenberg, ed., *Capital, Efficiency and Growth*, Cambridge, Ballinger, forthcoming.

Gollop, Frank M., and Dale W. Jorgenson [1980], "U.S. Productivity Growth by Industry, 1947-1973," in John W. Kendrick and Beatrice M. Vaccara, eds., *New Developments in Productivity Measurement*, Chicago, University of Chicago Press.

Hicks, John R. [1932], *The Theory of Wages*, London, Macmillan (2nd edition, 1963).

Jorgenson, Dale W., and Barbara M. Fraumeni [1980], "Substitution and

Technical Change in Production," in Ernst R. Berndt and Barry Field, eds., *The Economics of Substitution in Production*, Cambridge, Ballinger, forthcoming.

Jorgenson, Dale W. and Zvi Griliches [1972], "Issues in Growth Accounting: A Reply to Edward F. Denison," *Survey of Current Business*, Vol. 52, No. 5, Part II, pp. 65-94.

Jorgenson, Dale W., and Lawrence J. Lau [1980], *Transcendental Logarithmic Production Functions*, Amsterdam, North-Holland, forthcoming.

Kendrick, John W. [1961], *Productivity Trends in the United States*, Princeton, Princeton University Press.

—— [1973], *Postwar Productivity Trends in the United States, 1948-1969*, New York, National Bureau of Economic Research.

Samuelson, Paul A. [1953], "Prices of Factors and Goods in General Equilibrium," *Review of Economic Studies*, Vol. 21, No. 1, October, pp. 1-20.

Chapter 11

Energy Output Coefficients: Complex Realities behind Simple Ratios

G.C. Watkins* and E.R. Berndt**

INTRODUCTION

The demand for energy is a derived demand in that it is transmitted from demands for goods and services which incorporate energy as an input. Trends in the ratio of energy consumption to the level of output—the so-called energy coefficient—are often used to examine energy demand in the industrial and other demand sectors. In a market economy, the inference of this approach is that at a time of increasing energy prices, a rise in the energy coefficient is an indication of waste and inefficiency or of a perverse price response. Correspondingly, a fall in the energy coefficient is evidence of the efficacy of the price mechanism and government regulations in enhancing energy conservation. Over the interval 1971–1976, in the Canadian textile industry there has been virtually no change in the energy:output coefficient, despite substantial increases in real energy prices. Casual analysis might suggest, therefore, that energy price elasticities are miniscule. This paper seeks to dispel such a myth.

Work undertaken in the 1970s, of which the study by Berndt

*DataMetrics Limited and University of Calgary.
**University of British Columbia.

and Wood is an early example, suggests that the use of trends in energy coefficients to analyse energy demand is too simplistic. Their work showed the importance of looking behind the energy-output coefficient to discern what is happening. Potentially important factors to explain variations in energy : economic activity ratios include the structure of the economy, energy prices, and prices of other inputs used in the production of goods and services. If these factors were significant, extrapolation of energy coefficients on some kind of trend basis would be appropriate only where relative energy prices and the structure of the economy remained constant. Both these hypotheses are tenuous. A more comprehensive technique of analysis is to employ the economic theory of production which treats energy as one input among others which in combination produce output. Typically, the other inputs in addition to energy are physical capital, labour and non-energy intermediate materials.[1]

Thus, the methodology for investigating determinants of energy coefficients involves characterization of an economy or industry by a production and associated cost function which explicitly identifies energy as an input in the production of goods and services. This permits examination of what happens to energy output coefficients as economic incentives induce a variation in the input mix required to produce a given level of output. For example, suppose the price of labour increased at a rate faster than the price of energy. If energy and labour were substitutes, the relative increase in the price of labour to that of energy would tend to induce higher energy consumption by substituting relatively cheap energy for more expensive labour. But if energy and labour were complements, the higher price of labour would tend to reduce both the demand for labour and the demand for energy. The estimation of relationships between factors of production according to whether they are substitutes or complements is a prime feature of the production and cost function approach.

However, in such analysis it is important for the model to incorporate the long-lived nature of typical energy-using equipment. When this is recognized, short run energy price responses will generally tend to be smaller than long run impacts.

The main purpose of this study is to provide a comparative analysis of four models of energy demand which may be used to examine energy:output coefficients. The four models graduate from simple to complex specifications. To avoid obscuring model comparisons by

---

[1] Moreover, if the data permit, a breakdown of the capital stock into structures and machinery-equipment and labour into blue collar and white collar elements would be desirable.

possible aggregation problems, we concentrate our attention on one industry, Canadian textiles, 1957–1976.

[Full specifications of the models are available in the Full Proceedings Volume of the 1980 IAEE Conference.]

SALIENT CHARACTERISTICS OF THE CANADIAN TEXTILE INDUSTRY

There are over 600 textile mills throughout Canada. Approximately 30 percent of these mills are located in Ontario and 60 percent in Quebec. Over half of the eastern mills operate within the Toronto and Montreal areas; many of the textile mills function as industrial centres of small communities.

Canadian textile mills produce synthetic, woolen and knit fabrics for domestic use—such as clothing, linens and floor coverings—as well as automotive fabrics, asbestos products, paper makers supplies, fabrics for building products, canvas, cord and laminated fibres.

The Canadian textile industry is relatively inefficient by international standards; exports are negligible, while tariffs and quotas are imposed on imports. Nevertheless, compared with some other countries, Canadian import controls are relatively lax.[2] For instance, the United States and the E.E.C. countries restrict imports to a 20 percent share and 25 percent share of their respective domestic markets, while Canada allows imports to absorb almost half its domestic market. Most of Canada's imported clothing comes from the Far East.

Table 11.1 shows trends in input and output prices in the textile industry, 1957 to 1976. The price of capital and labour have both increased fourfold in nominal terms since 1957. Up until 1972, the price of energy was quite stable, but since then it has increased more rapidly than the price of any other input. The price of non-energy intermediate materials has shown a much lesser degree of variation than other inputs, and has increased at much the same rate as the implicit price of output.[3]

The quantity indices shown in Table 11.2 show a rather constant amount of labour over the 1957–1976 period—notwithstanding a

[2] Canada's import policy for clothing and other government policies affecting the domestic textile industry are in the process of being revised, ostensibly to provide a more favourable economic climate for domestic textile producers.

[3] This output price index is not a weighted average of the input prices, since it takes account of technological progress (cost diminution).

Table 11.1. Canadian textile industry: trends in prices (1971=1, nominal prices)

	P_K Capital	P_L Labour	P_E Energy	P_M Materials	P_Y Output
1957	0.4727	0.4684	1.107	1.081	1.042
1961	0.5849	0.5411	0.930	1.079	1.034
1966	0.8008	0.6804	0.913	1.066	1.046
1971	1.0000	1.0000	1.000	1.000	1.000
1972	1.1077	1.0696	1.005	1.001	1.006
1973	1.0620	1.1361	1.090	1.048	1.056
1974	1.3547	1.2753	1.402	1.232	1.241
1975	1.6350	1.5000	1.630	1.301	1.286
1976	1.6286	1.6962	1.890	1.386	1.380

Source: Based on data from Economic Council of Canada.

Table 11.2. Canadian textile industry: trends in quantities (1971=1).

	K Capital	L Labour	E Energy	M Materials	Y Output
1957	0.7449	0.9360	0.4902	0.3340	0.3649
1961	0.6921	0.9087	0.5795	0.4003	0.4419
1966	0.9171	1.0457	0.7931	0.6575	0.6818
1971	1.0000	1.0000	1.0000	1.0000	1.0000
1972	1.0286	1.0829	1.1736	1.1365	1.1306
1973	1.0688	1.1384	1.1707	1.2462	1.2267
1974	1.1160	1.1123	1.1884	1.2053	1.1925
1975	1.1578	0.9943	1.1989	1.1362	1.1314
1976	1.1505	0.9950	1.1628	1.1811	1.1649

Source: Based on data from Economic Council of Canada.

tripling of output. Materials seem to have increased essentially on a pro-rata basis with output. Energy consumption has increased at a lesser rate than overall output, but has kept pace with output since 1971, despite the increase in energy prices. The growth in capital employed has been at about half the rate of output growth 1976 over 1957, but has risen at much the same rate as output since 1971.

The cost shares of inputs in the textile industry have been reasonably stable over time. In 1976, the shares in gross output costs were:

Materials — 60 percent
Labour — 27 percent
Capital — 11 percent
Energy — 2 percent

Note the small share of energy; this is similar to the manufacturing sector in total and suggests that contrary to apparent government concerns, changes in energy prices do not have a strong impact on industrial costs.

CONCLUSION

We have examined energy:output ratios for the Canadian textile industry. Especially in the 1970s, at first glance the trend in these ratios seem bizarre. Notwithstanding higher energy prices, the energy:output ratio rises in some years, and the 1976 ratio is much the same as that in 1971. Prima facie, this suggests energy prices have had virtually no impact on energy consumption in this industry.

We have employed a number of economic models to explain these trends. The first was a naive model which simply treated energy demand as a function of energy prices and output. It was unable to account for fluctuations in the energy:output ratios in the 1970s. The second model was more complex and distinguished between short run and long run impacts, but it was also inadequate to explain the post 1970 trends. Neither of these models were grounded in any economic theory of production. The third model was predicated on such a theory, but was static in formulation, and again was not able to comprehend recent energy:output ratios. Much more encouraging results emerged from a dynamic version of the third model. Such a model was able to track the direction of movements in energy ratios and provide an economic rationale for them. Contrary to the apparent evidence, energy prices were found to have restrained energy consumption.

Our overall conclusion is that energy:output ratios are a useful summary statistic, but can be misleading. They cannot of themselves convey knowledge on the degree to which energy prices are affecting energy demand. Investigation of such effects requires more complex economic models to look behind energy:output ratios and disclose the milieu of determining relationships. Moreover, such models should allow for a distinction between short run and longer run impacts.

REFERENCES

1. Balestra, P. and M. Nerlove, "Pooling Cross Section and Time Series Data in the Estimation of a Dynamic Model: The Demand for Natural Gas", *Econometrica*, Vol. 34, No. 3, July 1966, pp. 585-612.
2. Berndt, E.R., M.A. Fuss and Leonard Waverman, "Empirical Analysis of Dynamic Adjustment Models of the Demand for Energy in U.S. Manufacturing Industries, 1947-74", Final Report, Palo Alto: Electric Power Research Institute, 1979.
3. Berndt, E.R., C.J. Morrison and G.C. Watkins, "Dynamic Models of Energy Demand: An Assessment and Comparison", in E.R. Berndt and Barry Field, editors, *Modeling and Measuring Natural Resource Substitution*, Cambridge, MIT Press, 1981.
4. Berndt, E.R. and G.C. Watkins, "Demand for Natural Gas: Residential and Commercial Markets in Ontario and British Columbia", *Canadian Journal of Economics*, Vol. X, No. 1, February 1977, pp. 97-111.
5. Berndt, E.R. and David O. Wood, "Technology, Prices and the Derived Demand for Energy", *The Review of Economics and Statistics*, August 1975, pp. 59-68.
6. DataMetrics Limited, "Price Elasticity of Apartment, Commercial and Industrial Demand for Natural Gas in The Consumers' Gas Company Franchise Area", Calgary, November 1979.
7. Denny, Michael, M.A. Fuss and Leonard Waverman, *Energy and the Cost Structure of Canadian Manufacturing Industries*, Institute for Policy Analysis Technical Paper No. 12, University of Toronto, August 1979.
8. Kraft, John and A. Kraft, "On the Relationship Between Energy and GNP", *The Journal of Energy and Development*, Spring 1978, pp. 401-410.
9. "Forecasts: The OECD Area to 1990?", *Oil and Energy Trends*, August 1979, pp. 12-15.
10. Manne, Alan S., and Thomas F. Wilson, "The Econometric Experiment of 1973", Stanford University Energy Modeling Forum, Working Paper ENF4.5 (revised) May, 1979.
11. Sakbani, M.M. and John J. VanBelle, "The Non-OPEC Oil Supply and Implications for OPEC's Control of the Market", *The Journal of Energy and Development*, Autumn 1976, pp. 76-85.
12. Watkins, G.C., "Petroleum Prices, Inflation and Industrial Costs", Paper for Business Outlook 1980 Conference, The Conference Board in Canada, September 27, 1979.
13. W.J. Westaway Company Limited, *The 1979 Textile Manual*, Canadian Textile Journal Publishing Company Limited.
14. Wilson, Carroll J. (Project Director), *Energy: Global Prospects 1985-2000*. Report of the Workshop on Alternative Energy Strategies, Massachusetts Institute of Technology, New York: McGraw-Hill Book Company, 1977.

Chapter 12

Energy/Employment Policy Analysis: International Impacts of Alternative Energy Technologies

Arnold Packer* and Wilbur Steger**

INTRODUCTION

The authors have recently proposed that solutions to the Free World's struggle with "stagflation" should devote substantial attention to a long-range international high-technology investment program designed to alleviate the problems of low productivity, declining real investment, higher unemployment, accelerating inflation, and decelerating growth endemic to this global economic condition.[1]

We did not base this conclusion on an overall agreement as what is wrong or why it is happening: clearly, there is no such consensus. Most, however, would agree with what the solution would "look like"; i.e., a reversal of stagflation's symptoms: moderate but sustained and concerted growth of the world economy, not merely temporary departures from the affliction, as with recent U.S. resistance to recession, the soundness of the German currency, or improvements in short-term balance of payment positions. All agree that the

*Assistant Secretary for Policy, Evaluation and Research, U.S. Department of Labor.
**President, CONSAD Research Corporation.

99

solution mechanism—whatever form it takes—should ultimately lead to reduced unemployment, moderate uncertainty about exchange rates, and avoid overheating individual economies. In addition, sooner or later there would be a world-wide reduction in energy-related inflation and increased employment. The vicious cycle where high inflation leads public and private decision-makers to a policy of low growth and high unemployment would be reversed. Less unemployment, globally, would lead to less job and industry protectionism and more private investment. The combination of higher investment and less protectionism, it is agreed, would lead to higher productivity and greater real GNP growth, in developing as well as developed countries. Furthermore, since productivity and economic growth are the source of real wage increases, the average wage earner would find himself gaining in his battle with inflation, and become more willing to practice wage restraint. The debilitating stagflation cycles—both the real and the perceived—would be reversed and a stable prosperity would return.

The purpose of this Paper is to advance our understanding of what we need to know to pursue one or another course in developing strategies for dealing with stagflation—focusing on what we believe to be at the core of the problem: productivity, technology and its transfer, the international economic mechanism, jobs, and real income worldwide. We will focus on analytical approaches to an improved understanding, though not slighting behavioral and strictly empirical methods. And we begin at the beginning: our understanding of the characteristics of the economic malaise—preliminary to understanding primary causes.

FROM SYMPTOMS TO UNDERSTANDING

So much for stagflation's symptoms. In our original discussions,[2] we reasoned that, while specific solutions to stagflation may not be obvious, our understanding could be furthered if we examined the characteristics of stagflation issues and how these might influence and shape the solutions. We could not understand how employment relates to emerging energy technologies without looking at the bigger picture.

We observed, for example, that a primary characteristic of stagflation problems is that they are worldwide, not national. The several "crises" pose considerably more direct problems for the European nations than for the United States, and the developing countries both affect and are affected, considerably, by world stagflation. Any solutions which are purely "national" in nature and which do not comprehend the effects on other nations, and further rounds of international feedback, may not only fail: they may, in fact, do permanent harm to the situation.

We also noted that many economic sectors are contributing, to an extent. to the worsening stagflation. Steel, non-fuel minerals, freight transportation, autos and the finance industries, to name a few, have from time to time been the focus of public attention. Whether increased government intervention was warranted in these instances—be it protectionism, bail-out, or other forms of assistance— has not necessarily reached the "crisis" stage of energy; nevertheless, erratic performances, investment, profit, and employment in these and other sectors have been factors in feeding stagflation. Broad-scale economic solutions, we argue, must meaningfully deal with all these key sectors, not merely in reversing trends, but also in identifying appropriate roles for each which can help make them partners in the recovery efforts.

Consistent with the concept that many sectors need to be involved is the characteristic of stagflation which affects all factors of production. Private investment is often singled out as a primary victim or, in some cases, a leading culprit; however, other productive factors are no less "responsible" and/or victimized. "Productivity"—of all factors, not just labor—is advanced as a leading "cause and/or effect", depending on the stagflation expert, and no doubt is caught up in the cycle. Management, as well as capital ownership generally, is argued to evidence undue degrees of risk aversion which become more pronounced as stagflation persists. And landowners both benefit from the inflation and also make it more difficult to perform simple and traditional land assembly-siting functions. Stagflation solutions need to find an appropriate and rewarding role for each of the productive factor groupings. Only then, for example, will emerging technologies improve both employment prospects and real wages for the wider group of impacted sectors.

Three other characteristics are more "conceptual" but serve with the others in making solutions difficult. The first we might categorize is that of problem complexity, or the sheer "messiness" of the stagflation problem. It is so all-encompassing in the ways it affects all of our daily lives that it is difficult for government agencies and political leaders to cope with it. Criticize, yes; constructively manage, very difficult; lead, almost impossible. If the characteristics described above are true "leit-motifs", no single government agency, short of the heads of nations and world economic institutions, could encompass—within one framework—all the sectors, factors, and perspectives. Those familiar with how governments function can recognize stagflation as the classic example of a problem in political economy whose solution will not work—will not even emerge—unless those at the highest levels exercise strong and effective leadership. Clearly, this leadership must be another characteristic of stagflation's solution space. Furthermore, given the current "conservative" attitude of governments everywhere toward new programs—particularly budgets for these—it will take strong political leadership to initiate a new program, even if it is to "correct" stagflation. Certainly, one characteristic of the solution space is that it will likely involve little or no additional public dollars. Government help, support and services, yes: government and private sector cooperation and coordination, yes; but no, or at the most a few, dollars, and only if these promise big private dollar leverage.

Second, conceptually, is the characteristic of longevity. Most observers agree that the current levels of stagflation have been "in the making" for many years, probably decades. Tracing of the inflation/unemployment experience since World War II reveals how the economy has moved from a "44' and a "45" to "55" and "66" Rule during the late 1960s and early 1970s.[3] The long-term nature of this seemingly worsening economic position does not negate the possibility that a fairly abrupt change in the world economy occurred in 1973. No one can doubt that the high inflation rates caused by the fourfold increase in the cost of energy and the "disaster" in food prices, coupled with a worldwide overheating and a rise in commodity prices in 1974, have contributed substantially to the current problem. Furthermore, flexible exchange rates, helpful in the long-run, have added to the short-term uncertainty. Some observers believe that the high degree of uncertainty have left individuals and countries unable to cope rationally with their new economic environment. All of this notwithstanding, few would doubt that the current problems are "secular", transcending the events of 1973. Longstanding fundamental trends have worsened and basic solutions need to be found, solutions which transcend the so-called "energy" or "food" crises popularized in recent years. (We recognize, of course, that both the "energy" and "food" problems have long-term antecedents, also).

Most relevant to the analytical concerns of our Paper, perhaps, is that characteristic of stagflation which makes it difficult to characterize, to conceptualize, to define, to encompass. Without an appropriate understanding, we have taken to symptom-fighting —and the indirect effects have often been the loser for it. Many, today, contend that the price and wage controls of the early 1970s, well-meaning as these might have been, set off an unintended set of expectation dynamics which only served to hasten stagflation. Economists, for example, have discovered many of their beliefs to be conceptually deficient, if not downright barren, in the light of events. The division into warring camps of market-followers (and regulation removers) versus the appropriate-intervenors (and public-private managers) is frequently only a red-herring in the solution space. The solution to stagflation will not be a dramatic return to the market-place of the late nineteenth century—even if such were one of the possible solutions strongly advocated. Instead, a characteristic of the solution space must include the development of a set of concepts with which most theoreticians and practitioners of "political economy" feel comfortable, can articulate, and use in the battle against stagflation.

It is this last objective which we take as the major purpose of improved analytic approaches. To develop a set of management, institutional, and economic concepts which can assist in the design of an international program to cure stagflation, we need a much improved conceptual understanding of the basic forces which, acting together, serve both to produce—and properly cope with—a reduced stagflation.

CHARACTERISTICS OF A WORKING SOLUTION

The ingredients of an appropriate solution must be tied to how we characterize the stagflation problem. At this writing, these would seem to be characteristics of an overall solution to stagflation:

1. Global. To match the global reach of stagflation, its "cure" would seem best promoted through an internationally coordinated approach. This approach need not be detailed or comprehensive planning, but, rather, a push in what are obvious "right directions" to simultaneously increase employment, productivity, investment, and to moderate inflation. The approach must recognize the relationships between the OECD world and the developing countries, both in terms of imports and immigration. Common problems quickly emerge: the U.S. problem with illegal immigration from Latin America is duplicated elsewhere, for example, by the French problem with immigrants from North Africa. There is no way that the wealthy countries can withstand the drive of the poorer countries to move towards income parity, nor should they.

2. Inputs to Foreign Economic Policy. Some would characterize this global-wide concern as calling for an "improved foreign economic policy", encompassing government decisions and actions affecting both foreign and economic concerns.[4] To assist in these policy determinations, coordinated decisions and actions would appear to involve both: (a) the management of policy decision process so that trade-offs among policy interests and goals are recognized, analyzed, and presented to the president and other senior executives before they make a decision; and (b) the oversight of official actions, especially those that follow major high-level decisions, so that these actions reflect the balance among policy goals that responsible officials decide upon.[5] Under the current administration of the coordination of national and international economic welfare, there is less of a tendency to give short shrift to foreign policy interests (although counter-examples might readily be given) than had been true in previous years. A statement made in the mid-1970s characterizes this concern for international technology development and transfer policy:[6]

"The overall recommendation of this report is that technology policy must be coupled with socioeconomic policy. At all levels of policymaking and across the broad spectrum of government activities, technological options and user-needs (or market-demand) must be brought together and integrated in policy-making. Such a recommendation seems self-evident and easy to accomplish. In fact, the coupling of these two aspects of government policy is too seldom achieved. While in theory it is easy to do, in practice it is exceptionally difficult because of institutional commitments and lack of sufficient knowledge.

3. "Key" Sectors: High Technology Energy Sectors. To further match the global nature of stagflation, the solution space needs to include mechanisms which encourage massive private investment in high productivity "key sectors" where increased output would produce the most substantial moderation of inflation. This is where investment in potentially high technology energy industries—heavy oils, tar sands, shale oil, unconventional gases, centralized solar, synthetic fuels, decentralized photovoltaics, and nuclear reprocessing and waste disposal—might make the most global sense, in terms of increasing U.S. technology exports while reducing global energy problems at the same time.

By high-productivity "key sectors" we mean that the developed countries would, by and large, adopt the development strategy most characterized by the Japanese selection process. Simply put, the

Japanese appear able to pick "winners" in the economic sweepstakes: they encourage these winners as they phase out industries that are under severe competitive pressure such as steel or shipbuilding. A global anti-stagnation strategy would be somewhat different in that each nation, taken separately and then together, would be creating winners best suited to its own comparative advantage, rather than imitating or trying to obtain benefits from winners established elsewhere. For analogies, it would be interesting to examine the U.S. postwar experience in developing high technology export products. Foremost on that list might be nuclear reactors, uranium enrichment, executive aircraft, computers, prefab construction, and microelectronics.[7] These products have not been generated by the free market completely unaided. Initial development has generally been supported by substantial government expenditures and other public aids. In many cases, of course, these expenditures were not directed towards the ultimate products but had other objectives such as defense or space exploration.

Involved in such decision-making processes, clearly, would be new and strengthened interactions between the public and private sectors, whether these activities concern foreign direct investment, exports, and imports, and/or technology licensing. Multinationals (MNCs) and both developed and developing government agencies, are increasingly discussing and negotiating incentives conducive to mutually beneficial private technology transfer, considering such factors as the users' capability to absorb the transfer, as well as the legal, political, economic, and cultural environments of both supplier and receiver countries.[8]

4. The Developing Nations. Further, internal development in the less developed worlds would be encouraged by this strategy. This would be especially true if that development takes the form of reducing pressures on scarce world supplies such as energy and food or reducing the cost of providing essential services such as health and education. Large scale investment of private funds with government support, we believe, should be encouraged to avoid the need to increase government expenditure. Such private investment would help the United States and other countries reduce their federal budget deficits, an issue with high political visibility throughout the world. The most advantageous situation would be to use stocks of funds held by such countries as Saudi Arabia, Kuwait, Germany, and Japan. It would clearly be less inflationary if these funds were used to finance productive investment if they are allowed to fuel speculative booms in land, housing, and commodities.

Furthermore, given the advent of the "New International Economic Order (NIEO)", there are new relationships emerging between the MNCs and less developed countries (LDCs).[9] Critically important for our purposes, here, is that foreign direct investment projects are no longer considered by the host governments in the LDCs without being integrated into the national economic development plans, the sectoral plans, and the regional plans. Both domestic and foreign investment projects are increasingly evaluated according to the economic and social objectives traced by those development plans. The narrowly defined private profit objective of the firm often fails to coincide with the more comprehensive development objectives of the LDCs. The traditional private investment evaluation process usd by the MNCs is therefore increasingly of limited usefulness when brought to the attention of the host governments in the LDCs for evaluation and acceptance.[10] Below, we shall see how this need to include "social pricing" of costs and benefits in foreign direct investment evaluation conditions the requirements for analytical technigues useful in this decision process.

5. What Happens to the More Traditional Industries? In the developed nations, as appropriate to each case, steel, autos, textiles, and traditional manufacturing would be weaned away from government assistance, except as these may be agreed upon for short-run equity and social reasons. We are only now beginning to recognize the strengths and weaknesses in Free World technology and what can be done to reduce "technology gaps" between the United States and other developed nations.

Given both the high-technology and traditional industry consequences of this line of reasoning, developed and developing countries alike would need to develop mutually beneficial ways to decide and select preferred ways to provide:

- National and international high-technology export aid programs;
- Incentives for targeted "key sector" capital formation;
- Encouragement of private investment, generally, by selectively reducing regulation;
- The provision of economic development services including information, finance, and technology support resources;
- The encouragement of technology transfer and high-payoff RD&D programs, including substantially increased international cooperation;
- High and "appropriate" technology research;

- Complementary public investment support; and
- Tax expenditure and more direct investment subsidies.

6. All Factors of Production. All factors of production, not merely private investment of capital, must play a role in the solution of stagflation. Labor in high technology countries would require and would receive expensive, sophisticated training and would be rewarded with high wage jobs. High technology importing nations, specializing in medium and low technology and investment in more conventional industries, would be in lower-wage jobs, though almost assuredly better-off than under continued stagflation, high unemployment, and eroding real wages. Furthermore, on the wages and jobs front, there would be less attention paid to intranational regional disparities and more encouragement of worker mobility to places of highest productivity and labor need.

* * *

Substantial payoff, it seems to us, awaits the private and political institutions which successfully assume the leadership role, at the correct rime with the right anti-stagflation concepts and techniques. Rewards could occur in the short term even if many of the substantial economic payoffs would take place in the longer-run. A political administration demonstrating leadership and capacity to grasp the problems facing the economy, advocating and implementing programs of the sort described in this section, would be perceived as bold and strong. It should be much more appealing politically than a continued negative view that there is nothing that can be done but wait until high unemployment reduces inflation, a thesis that needs to be fully discredited, if it is not already. We believe that, in order to properly conceive, design, develop and implement such a "grand" counter-stagflation strategy, it will be necessary to improve upon the "theoretical constructs" and analytical approaches required to better understand and explicate each and every step of the complex decision process which constitutes "foreign economic policy". To such theoretical and analytical improvements we turn in the next section.

CONCEPTUAL AND ANALYTICAL REQUIREMENTS, AND AVAILABLE METHODS

It has been only seven years since Richard Caves said, at a November 1973 colloquium on "The Effects of International Technology Transfers on the U.S. Economy":[11]

> "Regarding analytical models and concepts, I can only protest that many of those used to investigate technology transfers and related matters are highly myopic and potentially deceptive in their policy conclusions. Take, for instance, the vast amount of labor expended on investigating the effects of foreign direct investment on the U.S. balance of payments. It should be clear by now that *any* answer can be produced, depending on what one supposes is the alternative to the outflows of investment that have actually occurred."

We do not claim that, in seven years, dramatic changes in methodology have taken place, nor that what has been accomplished since then did not draw upon that present in 1973—because, of course, the achievements have drawn upon this background. Indeed, there was much to build upon even then as Caves and others[12] in the early-to mid-1970s were clearly aware, including these types of studies and approaches:

1. Economic welfare significance of trade and investment policy, i.e., what general equilibrium international trade theory reveals about technology transfer in its pure form;[13]
2. Effects of direct investment on the balance of payments and foreign exchange/earnings positions of importing and exporting nations;[14]
3. Empirical, in-depth, and case studies of the behavior and activities of parties interested and involved in direct investment, licensing and/or other technology development and transfer activities;[15]
4. Employment, income, and growth effects of foreign direct investment, and/or trade;[16] and
5. Multinational, multisectoral input-output and/or econometric models of the world economy, both of the international economic variables and key international trade activities between nations.[17]

Another way of looking at what we do and do not know about the effects of international technology transfer and development is depicted in Figure 12.1, where the spectrum of international economic analysis methods is portrayed, ranging from very detailed simulations (possibly one-to-one iconic simulations of investor or multinational corporate behavior) and detailed behavioral studies, to empirical descriptive studies, then to partial and general equilibrium system models (both static and dynamic). The techniques to the right side, clearly: are more macro in purpose; are typically considerably more complex and "analytical" and aggregative in nature; represent a top-down approach to estimating conditional forecasts of individual nations, sectors, and key international economic activity variables; and tend to be more useful in "normative" policy contexts. Those to the left side: are micro in purpose; attempt to capture the cultural/psychological realities of the behavioral micro-units involved; are typically more simple to comprehend; are generally more descriptive; and represent a bottoms-up approach to forecasting and analysis.

While there are representative studies and models to illustrate each of these methods, the techniques tend to be relatively unevaluated relative to day-to-day uses in public sector decision-making and policy making. The last five years have seen a rapid growth in the development and use of "global" econometric and interindustry modeling, on the one hand, and detailed microsimulation of investment decision-making and multinational corporate behavior, on the other hand. Rapid advance is being made—but, of course, much remains to be done, particularly in terms of integrating these analytic approaches to improved understanding, public policy and management.

ASSESSMENT OF THE CURRENT
STATE-OF-THE-ART

We would anticipate that, in the next five to ten years, formulations will be developed which will combine the better and more policy-relevant and useful features of the methods listed in Figure 12.1: results of descriptive and behavioral studies will find "homes" as parameters in partial and general equilibrium models, and these models will become more capable of dealing with dynamic phenomena. Perhaps what is needed most crucially and rapidly is a better linkage of the international modeling to that of the national models and even the intranational models detailing the sectoral, spatial, and technological makeup of each nation. For example, while the problem of sectorally-balanced versus unbalanced development has been addressed for some time (Rodan, 1943, and Hirschman, 1958),[18] intranational spatial balance or imbalance received little

	The Real World Behavioral Studies	Empirical (Descriptive) Studies	Partial Equilibrium Models: Static	Partial Equilibrium Models: Dynamic	General Equilibrium Models: Static	General Equilibrium Models: Dynamic
Macro						
Micro						
Cultural/ Psychological						
Representation of Reality						
Complexity						
Cost to Develop and Use						
Bottoms-Up						
Top-Down						
Normative Uses						

Figure 12.1. Spectrum of international economic analysis techniques.*

*This is an adaptation from M.A. Geisler and W.A. Steger, "The Combination of Alternative Reserve Techniques in Systems Analysis", *Management Technology*, May 1963.

attention until the 1960s (Perloff, et. al., 1960; Boudeville, 1966; and Friedman, 1966).[19] Until recently, in fact, spatial or locational analysis of any given national economic plan was pretty limited. Economic planning in the 1950s and early 1960s focused on capital use and did not explicitly address land use, locational analysis, or subnational economies more generally (Harr, et. al., 1968).[20] The implicit, but untenable, assumption during these years was that if an appropriate allocation of capital was obtained, appropriate spatial allocation of labor, management and land would "naturally" follow suite.

Productivity. Intranational detail and structural validity would seem to be an absolute necessity for improved international economic policy analysis and decision-making about technology development and transfer, particularly as these affect (and are affected by) behavioral, cultural, and experiential differences between nations. A case in point is the relationship of technology development and transfer, on the one hand, and productivity change on the other. "Productivity" is a complex phenomenon and few, if any, models capture these following nuances in the sources of productivity change:[21]

> "This term "process improvement" (PI) describes this development towards the more efficient production of essentially the same product. "Process improvement", however, should be understood to stem only from nonstructural productivity increasing measures such as increases in the quality and quantity of the capital stock or managerial capacity, or in the quality of the labor force. Technological changes that radically alter the structure of the industry are considered to be a separate phenomenon, an example of which is the "supermarket revolution" of the 1950s and 1960s. . . ."

> "Changes in economic structure are also significant. An important part of this "structural" hypothesis is that a major cause of productivity growth is the introduction of either wholely new products or at least very different ways of performing the same functions. These productivity increases occur not because someone works harder, or uses more capital to produce the same kind of steel, but because someone moves from a low-productivity— and usually low-wage—activity to a higher productivity job."

We believe that, unless an analytic method captures these different aspects of productivity change, it will *not* be able to depict and represent what happens when a technological change is taking place. Most analysts would agree that the only legitimate reason for making a model more complex is to increase its "richness" and accuracy. A number of assessments have reviewed models in terms of the tradeoff between degrees of error (error of specification, and error of measurement) versus model complexity.[22]

Technology-Economics Data Bases. There are techniques to deal with real-world complexity when there is a compelling need to do so. Quite often, the virtues of greater realism and accuracy versus the increased costs of complexity and possibility of measurement error are decided in favor of regional (or subnational) in-depth knowledge. Aggregation problems leading to cover-up of crucial structural interrelationships can generate significant, difficult-to-correct and detect specification errors. Basically, this argument for national and subnational input in global modeling is tantamount to that time-honored phrase, "You've got to know the territory".

The other major element currently lacking in the major analytic, modeling approaches to the study of international technology development and transfer is the critical detailed translation of technological data to economic data—and then incorporating this knowledge into the relevant national and international frameworks.

One of the co-authors has been involved in translating detailed technological data about new, evolving energy technologies into models capable of estimating conditional employment and environmental impact assessments of alternative development scenarios for these technologies.[23] While, on the micro level, economic assessments of energy technologies have concentrated primarily on market penetration and commercialization feasibility issues, these assessments have provided only minor emphasis to the prospective employment implications of individual technologies as forecast in macro national models. For example, the direct and indirect employment effects associated with alternative energy technologies (e.g., eastern underground versus western surface coal, solar energy, synfuels, etc.) has received limited analytical attention at best. Such estimates, however, *are* perceived as critical information for both national and international employment, manpower and energy policy analysis. Specifically, a growing interest has emerged in the possible use of energy policy explicitly to create needed and meaningful employment opportunities. This is *crucial* in any consideration of technology development and transfer.[24] Appendix A to this paper provides a short description of the DOL-DOE energy-employment effort which utilizes the detailed engineering-economic technology base used in many of the DOE's decision-making and planning exercises.

Other Impacts. Similarly, national and international *environmental* consequences of new (energy) technologies are increasingly critical variables in policy analyses regarding technology development and transfer. Clearly, nations may place different values on reducing pollution and specific pollutant residuals. The United States may place a higher value on clean air and water, and may wish, in a tradeoff sense, to export both jobs and specific high-cost (social and private) environmental problems to a nation explicitly willing to receive both. This cannot be done—analytically at least—without the detailed knowledge of technology-economic-environmental relationships. For ecample, in the case of solar energy:[25]

"The use of more solar energy leads to both increases and decreases in residuals. On the one hand, solar energy generates its own brand of pollution including particulates from wood burning, biochemical oxygen demand and suspended solids from biomass farms, and pollution from the manufacture of solar equipment (solar indirect residuals). On the other hand, solar energy reduces some residuals. For example, it displaces other energy forms like coal and oil where direct residuals are relatively high. Also, as the manufacture of solar equipment increases (which, as noted, raises solar indirect residuals) the manufacture of other things declines leading to an offsetting influence on indirect residuals."

CONCLUDING REMARKS

Detailed employment and environmental consequences—intranationally, nationally, and internationally—are illustrative of the technology data bases needed for incorporation into global modeling purporting to offer analytical assistance to informed foreign economic policy analysis. We believe that, in the next several years, there will be effective, increasing recognition of these (and other) diverse requirements for such informed and constructive decision-making, as well as increased capabilities—both piecemeal and integrated—to meet these stringent requirements.

We do not pretend that these first steps are all that is needed to fully explicate a sure-fire, technology based, anti-stagflation program. But we do believe the above represents a necessary, if not sufficient, approach to learning how an international, technology-based anti-stagflation program can be devised and put together. These specifications need much more fleshing out, but it is time to begin the effort, in earnest and with a coordinated set of resources committed to the challenge.

[Full details of the Improvement of Employment/Energy Economic Data Bases and DDE/DOL Impact Analyses are included in the Full Proceedings Volume I.]

NOTES

1. W.A. Steger, "Designing an International High-Technology Investment Program: An Alternative to Global Stagflation", CONSAD Research Corporation, April 1979; and A. Packer, "The Alternative to Stagflation", U.S. Department of Labor, unpublished paper, March 1979.
2. *Ibid.*
3. This "rule" concerns the limits that, at any point in time, to which an economy can experience combinations of employment levels and rates of inflation.
4. "Foreign economic policy, for purposes here, includes government actions with important impact on U.S. relations with other governments and on the production and distribution of goods and services at home and abroad". I.M. Destler, *Making Foreign Economic Policy*, Brookings Institution, 1980, p. 7.
5. *Ibid.*, p. 8.
6. Subcommittee on Economic Growth, Joint Economic Committee, U.S. Congress, *Technology, Economic Growth, and International Competitiveness*, U.S. Government Printing Office, Washington, D.C., July 1975, p. 65.
7. In the future, a number of high-technology energy, food, health, education, pollution control, and sophisticated manufacturing industries could be added to this list of "winners".
8. Sumye Okubo, "Industrial Innovation and U.S. Policy: International Transitions", National Science Foundation, 1979; Public Policy and Technology Transfer Project, *Viewpoints of U.S. Business*, U.S. Department of State, 1978.
9. A code of conduct for the MNCs has been considered and formulated by the United Nations Center on Transnational Corporations: "Transnational Corporations: Texts Relevant to an Annotated Outline Suggested by the Chairman of the Intergovernmental Working Group on the Code of Conduct", 26 January 1978. See also: International Chamber of Commerce, "Guidelines for International Investment' Proposal of the ICC", Paris, 29 November 1972; International Confederation of Free Trade Unions, "Charter on Multinational Enterprises", Mexico, 1975. The OECD Countries also adopted the "Declaration and Guidelines on International Investment and Multinational Enterprises" in mid-1976.
10. A.D. Cao, "Social Pricing in Foreign Direct Investment Evaluation", *Business Economics*, January 1980, pp. 53–60. Also: Harvey W. Johnson, *An Era of Mutual Progress: The Interdependence of U.S. Multinational Companies and Latin American Host Nations*, United Brands Company, Boston, Mass.; Richard J. Barnet and Ronald E. Muller, *Global Reach—The Power of the Multinational Corporations*, Simon and Schuster, New York, 1974; and I.M.D. Little and J.A. Mirrlees, *Project Appraisal and Planning for Developing Countries*, Basic Books, New York, 1974.
11. Richard Caves, "Effects of International Technology Transfers on the U.S.

Economy", pp. 29-42, in the National Science Foundation, *Papers and Proceedings of a Colloquium, The Effects of the International Technology Transfers on U.S. Economy*, July 1974.

12. Robert B. Stobaugh, "A Summary and Assessment of Research Findings on U.S. International Transactions Involving Technology Transfers", Gary C. Hufbauer, "Technology Transfers and the American Economy", Keith Pavitt, " 'International' Technology and the U.S. Economy: Is There a Problem?", in National Science Foundation, *Papers and Proceedings, ibid.*

13. "These models do not easily yield operational research designs, but they do provide a framework that is highly useful for exposing the gaps in more partial approaches. When technology is transferred without cost from country A to country B, the set of production possibilities is expanded in B and stays unchanged in A. In a two country model, the expansion of B's real income tends to improve A's terms of trade and welfare, but substitution effects associated with any sectoral bias of the technology transfer can cut either direction; unless there are complications due to consumption patterns, transfer of technology to B's export industry tends to improve A's terms of trade, and transfer of technology to B's import-competing industry tends to worsen them." (Caves, op. cit., p. 39). Also: N.S. Fieleke, *The Welfare Effects of Controls Over Capital Exports from the United States*, Essays in International Finance, No. 82 (Princeton, NJ: International Finance Section, Princeton University, 1971); Harry G. Johnson, "Economic Expansion and International Trade", *Manchester School of Economics and Social Studies* 23, May 1955, pp. 95-112; Harry G. Johnson, "Increasing Productivity, Income-Price Trends, and the Trade Balance", *Economic Journal* 64, September 1954, pp. 462-485; R.W. Jones, "The Role of Technology in the Theory of International Trade", in Vernon (ed.), *The Technology Factor in International Trade*, Columbia University, National Bureau of Economic Research, 1970, pp. 73-92; R.W. Klein, "A Dynamic Theory of Comparative Advantage", *American Economic Review* 69, March 1973; A.R. Mundell, "International Trade and Factor Mobility", *American Economic Review* 47, June 1957; R.E. Jones, "International Capital Movements and the Theory of Tariffs and Trade", *Quarterly Journal of Economics* 81, February 1967, pp. 1-38.

14. G.C. Hufbauer and F.M. Adler, *Overseas Manufacturing Investment and the Balance of Payments*, Tax Policy Research Study No. 1 of the U.S. Treasury Department (Washington, D.C., Superintendent of Documents, 1968); William F. Samuelson, Appendix B of Stobaugh, Iacuelli, Kirby, Samuelson, and Warren, *The Effect on the U.S. Economy of Eliminating the Deferral of U.S. Income Tax on Foreign Earnings*, Cambridge, 1973; G.G. Moffat, "The Foreign Ownership and Balance-of-Payments Effects of Direct Investment from Abroad", *Australian Economic Papers* 6, June 1967, pp. 1-24; N. Bruck and F. Lees, *Foreign Investment, Capital and the Balance of Payments*, New York University, *Institute of Finance Bulletin*, 48-49 (New York, 1968); P.H. Lindert, "The Payments Impact of Foreign Investment Controls", *Journal of Finance* 25, December 1970, pp. 1083-1099; C. Hui and F.G. Hawkins, "Foreign Direct Investment and the United States Balance of Payments: A Cross Section Model", *American Statistical Association Proceedings of the Business and Economic Statistics Section*, 1972, pp. 21-28.

15. Stobaugh, *The Product Life Cycle, U.S. Exports, and International Investment*, Basic Books, 1971; Raymond Vernon, *U.S. Controls on foreign*

Direct Investments—A Reevaluation (New York, Financial Executive Research Foundation, 1969), pp. 39-64; Robert B. Stobaugh, et. al., "U.S. Multinational Enterprises and the U.S. Economy", Part II of U.S. Department of Commerce, *The Multinational Corporation: Studies on U.S. Foreign Investment*, Volume I (Washington, D.C., Superintendent of Documents, 1972); Piero Telesio and Jose de la Torre, "The Effect of U.S. Foreign Direct Investment in Manufacturing on the U.S. Balance of Payments, U.S. Employment and Changes in Skill Composition of Employment", *Occasional Paper No. 4*, Center for Multinational Studies (Washington, D.C., February 1972); J.F. Gaston, "Why Industry Invests Abroad", *The Multinational Corporation: Studies in U.S. Foreign Investment, Volume 2*, (Washington, D.C., U.S. Government Printing Office, 1973); Robert D. Stobaugh, Dario Iacuelli, John C. Kirby, William F. Samuelson, and Theodore R. Warren, *The Effect on the U.S. Economy of Eliminating the Deferral of U.S. Income Tax on Foreign Earnings*, op. cit.; S. Hymer, *The International Operation of National Firms: A Study in Direct Foreign Investment*, unpublished Ph.D. dissertation, M.I.T., 1960; "The Efficiency (Contradictions) of Multinational Corporations", *American Economic Association Papers and Proceedings* 50, May 1980, pp. 441-448; "The Internationalization of Capital", *Journal of Economic Issues* 6, March 1972, pp. 91-111; F.T. Knickerbocker, *Oligopolistic Reaction and Multinational Enterprise*, (Boston, Harvard Graduate School of Business, 1973).

16. "There is no analytical model, not even a static one, that relates U.S. employment to U.S. foreign direct investment. This is an area of model-building in which research should be encouraged." (Stobaugh, "Summary and Assessment of Research Findings", op. cit., p. 24); also: Jose de la Torre, Jr., Robert B. Stobaugh, and Piero Telesio, "The Effect of U.S. Foreign Direct Investment in Manufacturing on Changes in Skill Composition of U.S. Employment", in Duane Kujawa (ed.), *American Labor and the Multinational Firm* (New York, Praeger Publishers, 1973), also reprinted as Part II of *Occasional Paper No. 4*, Center for Multinational Studies (Washington, D.C., February 1973); Donald B. Deesing, Peter B. Kenen, Helen Waehrer, and Merle I. Yahr, contributors to Peter B. Kenen and Roger Lawrence, *The Open Economy: Essays on International Trade and Finance* (New York, Columbia University Press, 1968); R.;G; Hawkins, "Job Displacement and the Multinational Firm: A Methodological Review", *Occasional Paper No. 3* (Washington, D.C., Center for Multinational Studies, June 1972); R.B. Stobaugh with P. Telesio and J. de la Torre, "The Effect of U.S. Foreign Direct Investment in Manufacturing on the U.S. Balance of Payments, U.S. Employment and Changes in the Skill Composition of Employment", op. cit.

17. The best known are the Ann Carter-Wassily Leontief model for the United Nations, and Lawrence Klein's Project LINK-World Economy Study, WEFA model, for the forecast of world trade and financial flows, and the analysis "of the nature and implications of global economic interdependence". See John Sawyer (ed.), *Modeling the International Transmission Mechanism*, (North Holland Publishing Company, 1979).

18. P. Rodan, N. Rosenstein (1957), *Notes on the Theory of the Big Push*, Center for International Studies, M.I.T. Press; Albert O. Hirschman (1958), *Strategy of Economic Development*, New Haven, Connecticut: Yale University Press.

19. H.S. Perloff (with J. Friedman) (1957), "Education and Research in Planning: A Review of the University of Chicago Experiment", In H.S. Perloff, *Education for Planning: City, State and Regional*, Baltimore, Md.: Johns Hopkins Press; J.R. Boudeville (1966), *Problems of Regional Economic Planning*, Chicago: Aldine Publishing Company; and J. Friedman (1966), "Planning as Innovation", *Journal of American Institute of Planners*, Vol. 32.
20. C. Harr, B. Higgins, and L. Rodwin (1958), "Economic and Physical Planning: Coordination in Developing Areas", *Journal of the American Institute of Planners*, Vol. 24, pp. 167-173.
21. Arnold H. Packer and Brian P. Brosnahan, "The Productivity Puzzle, or the Hounds That Wouldn't Bark", unpublished, November 1979.
22. R.M. Rauner and W.A. Steger, "Simulation and Long-Range Planning for Resource Allocation", *Quarterly Journal of Economics*, September 1962; M.A. Geisler and W.A. Steger, "The Combination of Alternative Research Techniques in Systems Analysis", *Management Technology*, op. cit.; D. Leinweber, "Essay on Data Quality, Model Complexity and Uncertainty", RAND Paper, presented at Proceedings of the Workshop on National/ Regional Energy Environmental Modelling Concepts, Reston, Virginia, May 1979.
23. CONSAD Research Corporation, *Final Report for the Joint DOE-DOL Project to Improve Employment/Energy Economic Analysis*, U.S. Department of Labor, November 1979; also, N. Dossani, et. al., "Solar Energy: Its Economic and Environmental Consequences", presented at the American Institute of Chemical Engineers Meeting, April 1979.
24. The employment analyses of technology transfer referred to above (p. 13) typically have not utilized the detailed engineering-economic data bases being developed in the high technology agencies such as the U.S. Department of Energy.
25. Dossani, "Solar Energy: Its Economic and Environmental Consequences", op. cit., p. 23.

PART VI

New Insights into Energy Demand

Chapter 13

The Route to Increased Efficiency

A. F. Beijdorff*

The crude oil price over most of this century (that is up to 1970) never exceeded $3/bbl in money of the day. This implied that from 1948-1970 the crude oil price decreased steadily in real terms. Most of the infrastructure currently in place in the OECD countries was designed with this comfortable prospect of falling real prices in mind. Since 1974, this has obviously changed (*see Figure 13.1*). It should be remembered that the technical potential for (further) energy efficiency improvements is very substantial and that much of the existing stock of buildings and industrial plant can still successfully and gainfully be improved.

It is generally believed that the "elasticity" of oil consumption is very low. This appendix attempts to illustrate the author's notion that this elasticity is not very low after all. It should be said beforehand, that the material shown is not conclusive proof, and much of the assertions made in this essay still have to rely on "hunch" and "notion" techniques.

If we trace the impact of the ten fold crude oil price increase (in real terms) between 1970 and 1980 on the final consumer selling

*Group Planning Co-ordination, Shell International Petroleum Company, Shell Centre, London, U.K.

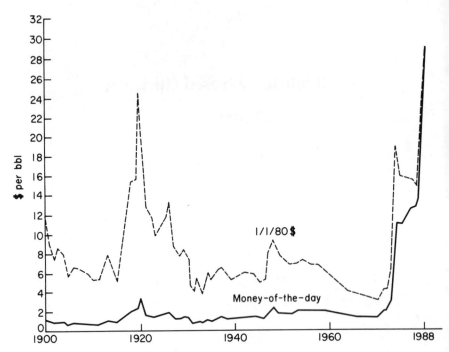

Since the turn of this century until 1970 the average crude oil selling price has ranged from $0.5 to 3/bbl in terms of money of the day. This implied, that, since 1948, crude oil became steadily cheaper in real terms. Most of the current oil consuming equipment was designed with this prospect in mind. Since 1970, when oil provided more than half of the total world primary energy, its price has gone up by roughly a factor of 25 in terms of money of the day, that is about 10 in real terms. With two price shocks behind us, we can now wonder what the demand elasticity for oil is.

Figure 13.1. Crude oil prices: an historical profile.

price, then we see that the selling price, in real terms, increased by a much lower factor than ten—for instance in the EEC by only a factor of roughly two across the barrel.

We should now distinguish two main components of the price mechanism: the income effect and the price effect (*see Figure 13.2*).

As the price of imported crude oil shoots up, direct effects (such as loss of spending power in the non oil sector, recycling delays) and indirect effects (consumer's increased savings ratio, loss of business confidence, a switch in government policies towards deflationary measures) will cause a recessionary slow-down in economic growth of consuming countries. This results in a lower economic output than otherwise would have resulted. This is referred to as the income effect. One can estimate this impact: some detailed studies suggest

PRICE works through	effects	effective cost S/boe
INCOME EFFECT	GDP (an increase in the price of imported crude oil has a recessionary effect on GDP growth)	250--500
PRICE EFFECT • Substitution	Oil/GDP • Coal/Nuclear in electricity generation and under boiler uses • Gas in Industry/Domestic	10–20 20–50
• Conservation — Improved Energy Efficiency — Improved Energy Management	• Transport • Domestic • Industry	30–60 20–50 5–60 0–50

Crude oil price increases work through to oil demand at two levels: income effects (expressed as a reduction in GDP growth) and price effects (expressed as a reduction in oil/GDP). The latter works through substitution and conservation, which in its turn can be subdivided into technological energy efficiency improvements (e.g. home insulation) and economies (i.e. improved heat management). The estimated loss of economic output (due to the income effect) divided by the reduction in oil demand due to the economic recession gives a measure of the overall economic cost of saving oil by economic deflation (i.e. $250–500/bbl). The costs of alternatives and conservation are considerably less.

Figure 13.2. The mechanism of crude oil price impact on oil demand.

that every additional $1 crude oil price increase reduces economic growth in the World Outside Centrally Planned Economies (CPE) by 1/4 to 1/2% over the years to follow.

Thus, an increase in the price of imported crude oil causes a loss in economic output. This in its turn causes a reduction in demand for imported crude oil. One can divide the loss in economic output (in billions of US $) by the amount of oil saved (in billions of barrels) and thus arrive at the implied cost of oil saved through economic recession. This illustrates that imported crude oil has a very high "nuisance value", in the order of $250–500 bbl.

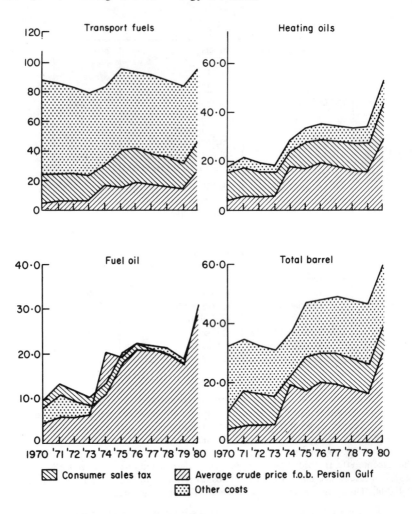

The crude oil price increases of 1974 hit the consumer in a much dampened way, due to the other price elements in the final selling price. Shown are (for Transport Fuels, Heating Oil, Fuel Oil and Total Barrel) the listed final consumer price in the EEC in constant 1980 money, broken down into: average crude price (f.o.b. Persian Gulf), sales tax (excise and VAT) and (by difference) all other costs. From this it is clear that on balance the motorist saw very little real price increase (as a result of the sales tax erosion in real terms), whereas heating oil went up by some 70% and fuel oil by some 90% in real terms. In contrast to 1970, where the share of the crude price in the final average selling price was some 12% currently it is over 50%, so that further real price increases will hit the final consumer much more directly than in 1974.

Figure 13.3. Oil product prices: EEC. $/bbl 1/1/1980 prices and exchange rates; 1st of Jan. each year.

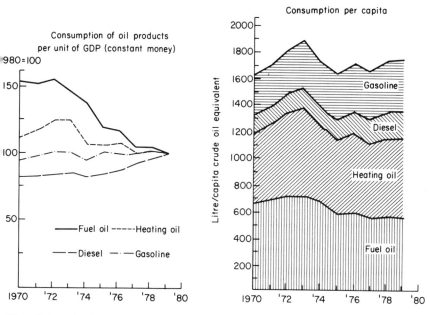

The relative price increases shown in Figure 13.3 can be traced to their effect on consumption. So as to eliminate the income effect, consumption of Oil Products is shown per unit of GDP (in constant terms). After steady increases over the sixties, 1974 was a turn-around point for all main products except diesel fuel. Fuel Oil/GDP came down by a third and heating oil by a quarter over just five years, whereas gasoline ceased to grow. This is in good qualitative agreement with the size of the corresponding relative price increases. (Diesel fuel increased its share since 1977 mainly due to the first time availability of a new range of diesel powered passenger cars and to taxation policies in some countries). Gasoline shows a marked plateauing in trend, (either as consumption per unit of private consumer expenditure or as consumption per household), a significant change, considering the small absolute increase in gasoline selling prices. On balance, the EEC now consumes less oil per capita than it did in 1973. Roughly half of the observed effects are due to substitution, the other half to a mixture of structural changes and conservation. These data suggest that the market reacts much faster to price signals than is commonly thought.

Figure 13.4. Recent trends in oil consumption in the EEC.

The second link in the price mechanism is the price effect, i.e. the reduction in consumption per constant unit of income. This is expressed by a decrease in oil/GDP for an increase in the oil price. It is interesting to note that the costs of substitution and conservation are significantly lower than the "nuisance value" indicated above. Although few energy consumers will realise this consciously, it is nevertheless an illustration of the strong incentive that is implicitly administered by the imported crude oil price.

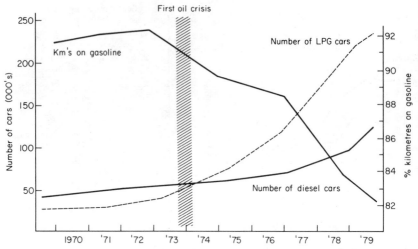

Another illustration that the market does react fast to clear signals. After the carless Sundays in 1973/1974 in the Netherlands, when drivers of LPG and diesel powered cars were exempt from the driving ban, interest in LPG powered cars increased sharply, not-with-standing an extra installation cost of roughly $900 per car. LPG is not taxed in the Netherlands, so that the selling price of a litre of LPG is roughly half that of gasoline. Diesel is taxed well below gasoline, with a resultant selling price roughly halfway between gasoline and LPG. It is interesting to note that gasoline cars lost some 8% of kilometres driven over just 6 years. The diesel cars only began to make an inroad after 1977, when for the first time a good range of diesel powered passenger cars came on the market.

Figure 13.5. An example of market reaction to shock & price signal. Recent displacement of gasoline in the Netherlands.

If we now examine how the crude oil price increases hit the consumer in the main oil consuming markets, then—again for the EEC—we see that the spread has not been even at all (*see Figure 13.3*). The transport sector saw hardly any increases in fuel prices, whereas the industrial sector saw a virtual doubling (in real terms) of the fuel oil price.

Focussing now on the price effect (as opposed to the income effect) we can examine the consumption of the main oil products per unit of GDP (*see Figure 13.4*). Qualitatively, we see—again in the EEC—a very good agreement between the reductions in consumption per unit of GDP (in constant money) and the corresponding increases in relative prices.

Without wanting to draw too strong conclusions from such a short time series, and recognising that roughly half of the reduction shown is substitution rather than conservation, we would like to submit this

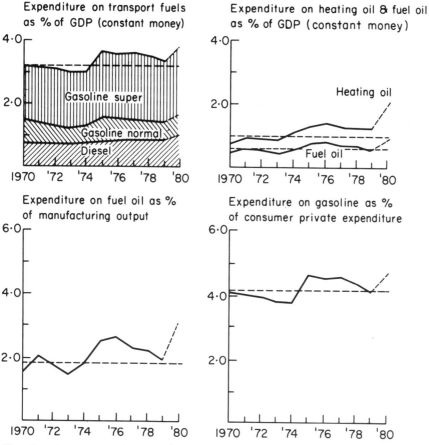

The oil product consumption data of Figure 13.4 can be combined with the price of Figure 13.3 to give the expenditure on oil products as part of total expenditure (or value added). One can observe, at least for transport fuels and for fuel oil, a trend of "constant percentage of income". If true, this would imply that the long term price elasticity of oil products is one, higher than commonly thought.

Figure 13.6. Recent trends in expenditure on oil as percentage of income in the EEC.

evidence as a nevertheless indication, that Adam Smith's "invisible hand" is still as active as ever, and still very effectively so.

A similar analysis (not presented here) has been carried out for the United States. There the price signals were not as clear as in the EEC, for well known reasons, but qualitatively, the same conclusions can be drawn. In real terms, 1979 was the first year that gasoline prices really went up significantly since 1968 or so in the United

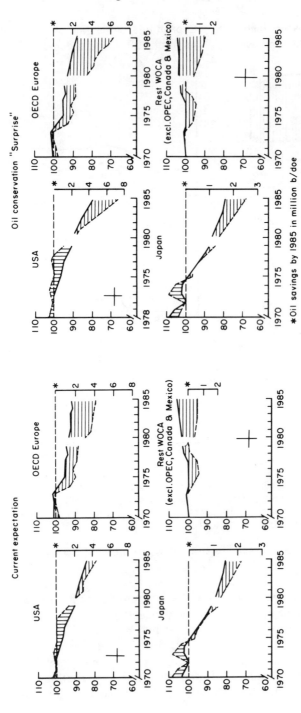

By plotting both total primary energy and oil per unit of GDP, indexes to the last pre-crisis reference year (1972), one can illustrate the overall reduction in energy and oil intensity, and, by difference, the amount of substitution by or for oil. USA, OECD Europe and Japan have all significantly reduced their energy intensity. Whereas OECD Europe and Japan had decreased their oil intensity by some 10–15%, the USA in 1978 was almost back where it was in 1972. Last year, however, saw a sharp turn-around which, given the certainty of oil price decontrol, now can be expected to continue if not accelerate. Given a conservative extrapolation of current trends and estimated availability of non-oil energy, the World outside Centrally Planned Economies would consume some 7 million bdoe of oil less than it would have on 1979 consumption patterns. One can foresee an extra effect due to the 1979 price shock; this could possibly reduce consumption by a further 6 million bdoe by 1985, given the same level of economic output.

Figure 13.7. Energy and oil consumption per unit of GDP (at 1970 prices).

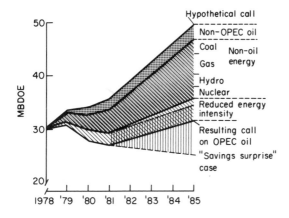

The relative significance of substitution and conservation are illustrated by their effect on a hypothetical call on OPEC oil (corresponding to an economic growth of some 2½% p.a. for 1979-1985 for world outside Centrally Planned Economies). If oil were the only balancing fuel, call on OPEC oil would increase to 50 million bdoe by 1985 (which would obviously not be met). With the anticipated net increases in non-OPEC oil production and the "conservative" extrapolation of energy intensities and availability of non oil energies of Fig.13.8, this call reduces to some 30 million bdoe by 1985. The "plausible oil conservation surprise" (or "the consumer gets the message" scenario) of Fig. 13.8, would reduce this call to some 25 million bdoe, which would ease the potentially tense oil supply situation considerably.

However, this "oil conservation surprise" would, on its own, not prevent the longer term global energy demand from increasing further, and a permanent solution will require concerted action and a determined effort in energy supplies.

Figure 13.8. A hypothetical reference case. Contribution of non-OPEC energy, and reduced energy intensity to reducing the call on OPEC oil.

States, and the reaction was clear: a 5% reduction in gasoline consumption in 1979, (i.e. a 7% reduction in consumption/GDP) over one year, in response to a real price increase 1978/1979 of not more than 30% (one has to discount this drop somewhat in the light of the actual shortages that occurred during the summer; however, mid 1980 demand was still down compared with 1978).

In 1978, US customers spent roughly 5% of total Consumer Private Expenditure on gasoline. With oil price deregulation now a firm fact one can predict that, in real terms, the gasoline selling price will go up by at least 60% and possibly as much as 100% as compared with 1978. This would imply that the percentage of private expenditure spent on gasoline would go up to 8-10%. However, at such a percentage level, the consumer will have to take notice, and the signs are everywhere loud and clear that the US motorist is taking notice.

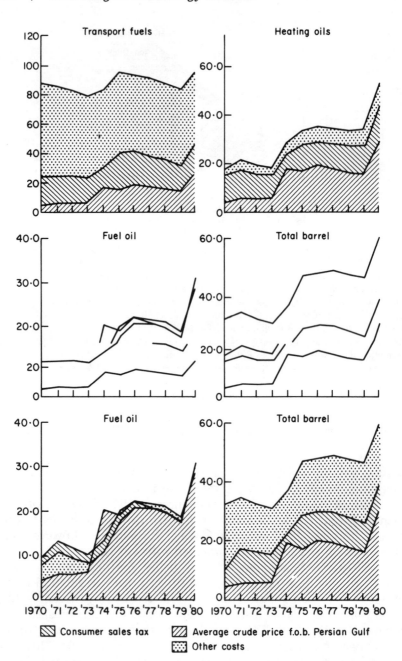

Figure 13.9. Oil product prices: EEC. $/bbl 1/1/1980 prices and exchange rates 1st of Jan. each year.

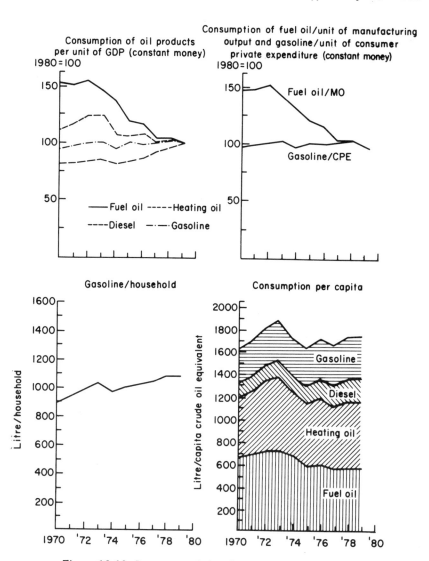

Figure 13.10. Recent trends in oil consumption in the EEC.

We can therefore now confidently predict that the US government's compulsory miles-per-gallon legislation will be able to boast some spectacular results over the next five years.

Another example of how price signals can work fast in the market place comes from the Netherlands. During the car-less Sundays in 1973 (as a result of the Arab oil embargo) vehicles powered by

Liquid Petroleum Gas (LPG) and diesel fuels were exempt. This made the public suddenly much more aware of the existence of such fuels. Added to that, the fact that LPG is not taxed in the Netherlands ensured a rapid penetration of LPG powered passenger cars. A loss of 10% in kilometres driven on gasoline over just 6 years is a very fast reaction indeed (*see Figure 13.5*).

Finally, one can examine the expenditure on fuels as a percentage of total expenditure. After all, if the price of a commodity doubles and your income doubles too (all in real terms of course) then you are not all that much worse off, in fact, none at all. Plotting expenditure on oil products as % of GDP, Consumer Private Expenditure, and Manufacturing Output in the EEC (*see Figure 13.6*), reveals a surprising tendency "back towards the norm", (a similar tendency can be observed in the United States). If it were indeed true that there are certain universal percentage ranges that consumers of different income levels tend to spend on certain categories of fuel consumption, then the corollary would be that the long term price elasticity of oil products is one, and that "long term" in this sense would mean only some five years or so.

The above observations, (together with many other signs) have led the authors to the notion, that over the next five years, simply due to the fact that "the consumer will get the message", the world could be in for a surprise on oil consumption. Rather than an inelastic demand that grinds inexorably into the medium term oil supply constraints, a scenario could be foreseen in which the world finds itself (perhaps to its own surprise) in a situation where the call on OPEC oil has fallen below 25 million barrels per day by 1985, whereas the GDP still has managed an average growth of some 2½% p.a. (*see Figures 13.7 and 13.8*).

Although this scenario still has all the chances of sliding into a crisis situation at any time over the next five years, it is interesting to note that it presents a plausible solution to the current dangerous situation (by allowing Saudi Arabia, for instance, to come down to more comfortable levels of production), and that the makings of this solution could already be emerging all around us. A careful monitoring of oil/GDP over 1980/81 will be a rewarding task for virtually all policy makers.

Chapter 14

Interfuel Substitution in European Countries*

*Frederic Romig and Professor Patrick O'Sullivan**

SUMMARY

A nation's fuel mix has been an important factor of its economic development and strongly affects its future options.

In the past, switching from coal to oil and gas assisted rapid growth. More extensive transport services, more efficient industrial processes and improved heating standards in buildings were achieved partly switching away from coal. Eastern European countries that still rely heavily on coal use twice as much energy per unit of GNP or NMP as western countries. But relatively efficient rapid growth has lead most western countries to large energy import requirements.

Future interfuel substitution to indigenous resources or to different fuels from more stable energy trading partners needs to take place much more rapidly than before. It should focus on the built environment. Unlike transport and some industrial processes, buildings can use any fuel for low temperature heat. The built environment can provide many options for the future if suitably designed and rehabilitated since it is the largest non-fuel specific sector in industrialised countries.

*This paper is based on the personal views of its authors and not the views of the institutions for which they work. Patrick O'Sullivan is Professor of Architectural Science at the University of Wales. Frederic Romig is Economic Affairs Advisor at the United Nations Economic Commission in Europe (ECE) in Geneva.

INTRODUCTION

A nation's fuel mix has been an important factor in its economic development and strongly affects its future options. Substituting one fuel for another has become a vital component of energy policy often necessary for the use of more indigenous resources, less imported oil, more renewables and the 'efficient' use of all available energy resources.

Most European countries are now diversifying both their energy trading partners and their fuel mixes. Diversified energy trading partners can give greater security of energy supplies in the short term. In the medium to long-term energy conservation coupled with interfuel substitution can keep vital options open.

Some economic sectors are more fuel specific than others. For example, barring a major break-through in battery technology, transport will continue to require liquid fuels. Some industrial processes such as glassmaking and ceramics also require particular fuels.

However, low-temperature heat mainly for use in building is the largest non fuel-specific sector in European countries. National energy options can be greatly enhanced by careful attention to the design and use of industrial, commercial, institutional, agricultural and residential building.

This paper examines some of the main issues of interfuel substitution in building. First, it takes a broad look at past trends. These show how total energy consumption has increased greatly in some countries since 1950 while it has risen far less in others. How Western European countries have substituted oil and gas for coal in buildings over the last twenty years. The resulting fuel mix for 1978 in the Western European buildings is given. This is compared with the 1976 fuel mix in four Eastern European countries.

Secondly, the paper looks at one country, the United Kingdom as an example of what can be done in others. It shows how fuels have been substituted one for another in the United Kingdom's built environment since 1950. It gives a detailed breakdown of 1978 energy consumption and points out that more than half the nation's energy goes to the built environment.

Thirdly, it examines some of the specific issues affecting interfuel substitution in buildings for the future. It explains how buildings can be assessed, designed, rehabilitated and managed so that different fuels can be used in them and looks at the lead times for such fuel substitution.

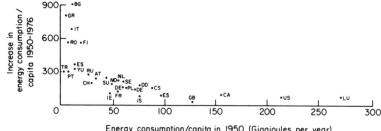

The following ISO country codes have been used: Austria: AT; Belgium: BE; Bulgaria: BG; Canada: CA; Czechoslovakia: CS; Denmark: DK; Finland: FI; France: FR; East Germany: DD; West Germany: DE; Greece: GR; Hungary: HU; Iceland: IS; Ireland: IE; Italy: IT; Luxembourg: LU; Netherlands: NL; Norway: NO; Poland: PL; Portugal: PT; Romania: RO; Spain: ES; Sweden: SE; Switzerland: CH; Turkey: TR; Union of Soviet Socialist Republics: SU; United Kingdom: GB; United States of America: US; and Yugoslavia: YU.

Figure 14.1. Increase in energy consumption per capita 1950-1976.

PAST TRENDS

Energy and economic development in European countries

Productivity and material standards have risen sharply (by any measure) in European countries during the last thirty years. Factories, roads, houses and railways—the energy-using infrastructure—were built up rapidly. Between 1950 and 1976 the length of railways for example has increased by 100 per cent. From 1960 to 1976 alone the dwelling stock has increased by 36 per cent and the length of roads by 10 per cent. Related activities and services increased apace. Iron and steel productivity rose by 77 per cent and automobile ownership by 150 per cent.

This unprecedented rise in energy consumption has been uneven among European countries. Certainly their usage in 1950 varied much more than it does now. Energy consumption per person was for example thirty times greater in the highest using nations than in the lowest then, whereas now it is only nine times as great. The maximum rate of increase has occurred in those countries designated low energy users in 1950 i.e. countries with an average per capita primary energy use of less than 60 GJ. Bulgarian energy use per capita rose more than nine times from 1950 to 1976. Italy also had a rise of well over six times and Romania one of nearly 5000 per cent. Conversely large energy users such as France, Poland and West

Germany increased by a relatively small 100 per cent. Figure 14.1 illustrates this development pattern. Heavier energy users generally increased less. Indeed, nearly half the total increase in energy consumption came from countries that were low energy users in 1950 even though they only took 25 per cent of 1950 energy consumption. This suggests that much of the rise in energy demand was 'needed' to acquire the energy-consuming infrastructure in use today and moreover, that as energy activity increased in some countries, increases in energy efficiency held down consumption in others. Certainly as car ownership increased overall, in many countries cars were on average driven less.

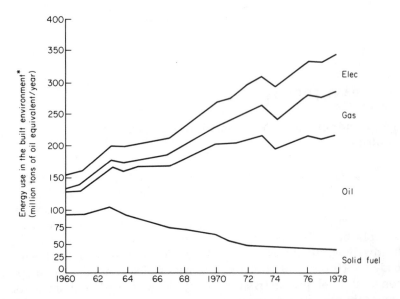

Figure 14.2. Built environment energy consumption by fuels in OECD Europe— 1960-1978.

Ultimately, such differences between countries can always be interpreted and explained in terms of climate, state of development, energy efficiency, size of domestic energy reserves and other factors. But this very large rise in energy use in some countries and small increase in others does raise the question of whether in the future, therefore, energy growth may slow down, or even flatten out, as population growth and the energy-using capital stock reach saturation in many countries.

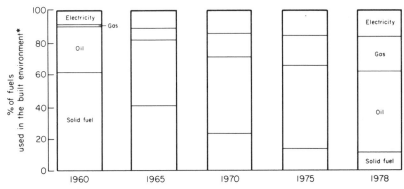

*Includes Agriculture which made up 5 per cent of the energy use shown here in 1978.

Figure 14.3. The fuel mix in the built environment* in OECD Europe 1960, 1965, 1970 and 1978.

Developing multifuel economies

European energy systems underwent other important transitions as the relevant countries' economies developed. In particular such countries used more energy per person and diversified the mix of fuels that they used. In 1950 many had virtually single fuel economies based on coal. Certainly coal accounted for 85 per cent of the 1950 European primary energy consumption, liquid fuels provided 24 per cent of the fuel mix while gas accounted for 11 per cent and the rest, about 2 per cent, came from hydro and nuclear electricity.

This switch in the fuel mix was closely linked to the speed with which many nations developed, as it provided the means by which they could build up new and better services more quickly. Oil and gas were used to provide more extensive transport services, improved home heating and more efficient industrial processes. Very large shifts in the energy using infrastructure occurred and fuel consumption shifted accordingly.

In 1950, the energy used in buildings in Western European countries was dominated by coal. In 1960 coal still provided 61 per cent of the delivered energy consumption (oil provided 24 per cent, gas 6 per cent and electricity 9 per cent).

From 1960 to 1978 the energy consumption in Western European buildings increased by 116 per cent to 345 M.T.O.E. During this period however per capita energy consumption increased by only 85 per cent because of increasing population.

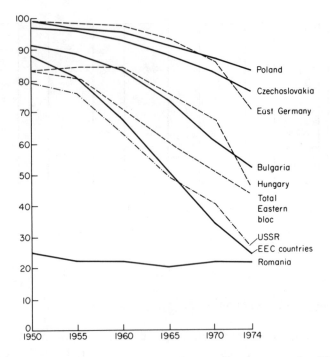

Figure 14.4. Solid fuel consumption in Eastern Europe countries 1950-1974 (percentage of total primary energy).

In the same period (1960 to 1978) the fuel mix changed considerably (*see Figure 14.2*) the proportion provided by coal reducing to 11 per cent, that by oil increased to 52 per cent, by gas to 21 per cent and by electricity to 17 per cent. Figure 14.3 shows the rate of this transition.

Eastern European countries

The fuel mix and its distribution by end use sectors has developed very differently in the Soviet Bloc over the last thirty years. Three important features stand out. First, the shift away from coal to oil and gas started much later and has progressed more slowly than in Western European countries (*see Figure 14.4*). Secondly, the Soviet Bloc countries use less energy in the transport sector (more goes to industry). Thirdly, total energy use is less efficient. Soviet Bloc total energy consumption per unit of GNP or NMP was nearly twice that of the EEC countries. The total delivered energy consumption

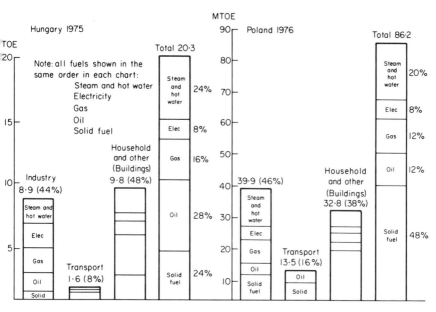

Figure 14.5. Delivered energy use in Eastern European countries 1975-1976.

per capita was nearly 50 per cent greater than in Western European countries.

Today Eastern European countries still use a greater proportion of coal because of the slower shift towards oil and gas during the 1950s and 1960s. As a result, their fuel mix in building differs very greatly from western countries. Coal use as a percentage of total primary energy is shown in Figure 14.4. A delivered energy breakdown is more revealing (*see Figures 14.5 and 14.6*).

The direct use of coal makes up over half of the energy consumed in the buildings of East Germany, Czechoslovakia and Poland. In addition, district heating and combined heat and power, mostly from coal, provide another large source of energy for buildings. Taken together, coal and steam and hot water provide 61 per cent of the energy used in buildings in Czechoslovakia, 61 per cent in East Germany, 50 per cent in Hungary and 78 per cent in Poland.

Delivered energy use per capita in the buildings of Eastern European countries is about 65 per cent of that in Western (OECD Europe) nations, but standards are very much lower. This is partly because fewer services are provided, houses and flats are smaller and energy is used less efficiently.

Although Eastern European countries use less delivered energy

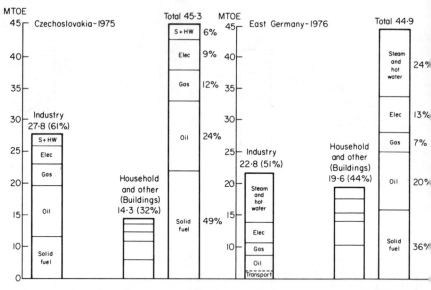

Figure 14.6. Delivered energy use in Eastern European countries 1975–1976.

per capita in buildings, they use more primary energy per person because much of their electricity generation and district heating is very inefficient. Specific fuel consumption of electricity generation measured in Kcal/kWh is 10 to 25 per cent higher in Eastern European nations than the average for the EEC. Energy consumption of the power stations themselves is 60 per cent higher in several Soviet Bloc countries. Transmission losses are between 60 and 115 per cent greater than in the EEC.

In Poland 66 per cent of all dwellings were flats in 1976 and the rest houses. In contrast, 18 per cent of United Kingdom dwellings are flats and some 72 per cent are terraced, semi-detached houses. Only 67 per cent of Polish houses have piped water, only 79 per cent have electricity supplies and only 44 per cent have fixed bath or shower facilities, while all United Kingdom houses have these as do most Western countries. In Poland, 20 per cent of dwellings have only one room, 35 per cent have two rooms and another 30 per cent have three rooms. Dwellings in the United Kingdom are larger with 30 per cent having four rooms, 34 per cent with five, 13 per cent with six and 7 per cent with seven or more.

One of the reasons for this disparity is that Soviet Bloc countries have devoted far more of their resources to industry. Indeed, many of their dwellings are flats built from industrial buildings components

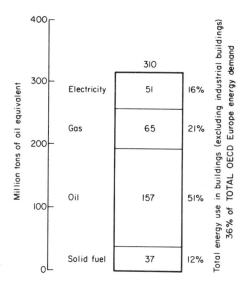

Figure 14.7. Energy use in commercial, institutional and residential buildings in OECD Europe—1977 (including industrial buildings).

which generally have higher air leakage and heat losses than do conventional masonry construction techniques.

The lower standards in Eastern European countries is partly a result of their fuel mix coupled with the relatively limited services provided. In Poland, for instance, only 38.7 per cent of dwellings were connected to the gas grid in 1978. A further 19.0 per cent of dwellings used bottled gas. In all, some 48 per cent of Polish towns had gas supplies in 1978. All towns had electricity supplies, but many rural dwellings did not. As mentioned above, only 80 per cent of dwellings were connected to the electricity grid in 1978.

Steam and hot water made up 20 per cent of Poland's national delivered energy use in 1976. About 22.7 per cent of total dwelling floor space had space and water heating from these sources.

In contrast, 70 per cent of United Kingdom dwellings had gas supplies in 1979. About 52 per cent of dwellings had central heating. Only 8 per cent of dwellings had solid fuel central heating, while 14 per cent had oil and 25 per cent had gas central heating. About 4 per cent had electric storage heaters, 20 per cent fan heaters, 27 per cent gas space heaters, 29 per cent liquid oil heaters and 70 per cent of dwellings had electric space heaters i.e. many United Kingdom dwellings already use more than one type of heating system and more than one type of fuel.

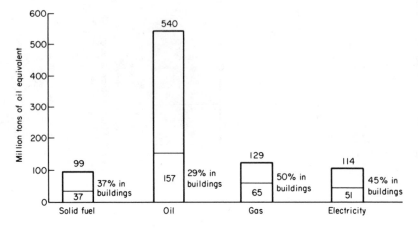

Source: OECD Energy Balances 1975–77.

Figure 14.8.

Western European countries

By 1977 the fuel mix for buildings in OECD Europe was far more dependant on oil than on any other energy source. Solid fuel had fallen to about 12 per cent of delivered energy use from 61 per cent in 1960, while oil use had more than doubled from 24 per cent to 51 per cent, gas 21 per cent and the rest, some 16 per cent, was electricity. Buildings also took a significant proportion of total OECD Europe delivered energy. In 1977, they accounted for 37 per cent of total delivered solid fuel, 29 per cent of oil, 50 per cent of gas and 45 per cent of electricity. The built environment's fuel mix and the share of fuels they used are illustrated in Figiures 14.7 and 14.8.

The United Kingdom

The delivered energy fuel mix for the United Kingdom's built environment has undergone a transition similar to that of western Europe as a whole, but moved rather more to gas than oil, as solid fuel use declined. The total delivered energy use did not rise greatly (see Figure 14.9). As shown in Figure 14.1, total primary energy usage was beginning to saturate in 1950. Indeed, primary energy use per household has been roughly constant since 1955, the well documented result of a decline in the use of solid fuel in inefficient

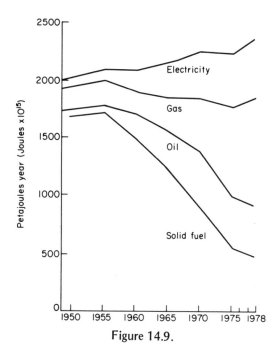

Figure 14.9.

open fireplaces and a rise in the use of more efficient oil and gas appliances.

The fuel mix changes shown in Figure 14.9 are for Housing, Commercial and Industrial buildings only. However, recent studies by the British Gas Corporation and the CBI suggest that much of the energy used in industry is in fact used in the buildings rather than in the process.

Buildings dominate energy use in the United Kingdom. In 1978, 57 per cent of the Nation's energy went into housing, commercial, institutional, and industrial buildings, at a cost of £8,000 million (Figure 14.10).

On a regional level, over half of OECD Europe's energy goes to the built environment if roughly the same calculation is done (*see Figure 14.11*). This means that at least 36 per cent and possibly as much as 52 per cent of delivered energy in Western European countries goes to heating, cooling, ventilating and providing hot water and services in buildings.

In other words, Western European countries have generally realised and maintained higher standards, and this has been based in part at least on inter-fuel substitution. Although these countries may

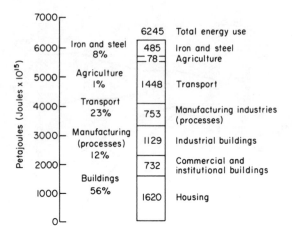

Figure 14.10.

now be paying a high price for this development in terms of oil dependency, they have as a result of this same development many options for the future. In particular the continued practice of inter-fuel substitution in the building sector, as part of a programme of energy conservation provides a way of maintaining the availability and flexibility of fuels to allow further development and change in those critical industrial, transport and other sectors.

The implications of this option are next explored in terms of the United Kingdom and thus by implication for all European countries.

FUEL SWITCHING IN BUILDINGS

Buildings are principally designed, built and operated by only six main groups of professionals and already have an existing mechanism for controlling the 'product' namely the Building Regula-tions. The majority of our buildings exist, such that the problem in the short to medium term is largely the 'realistic' one of management rather than one of prediction. In whole sectors of our built environ-ment the energy consumed is not simply related to GDP production. Further the energy is used for only four main purposes and nearly two-thirds of this usage is in the production of low grade heat. Since the production of this heat is not fuel-critical, buildings can in principle be used as a 'buffer' against any particular fuel shortages which may otherwise have a far greater impact in other sectors.

Buildings could, for example, be used to act as a buffer against

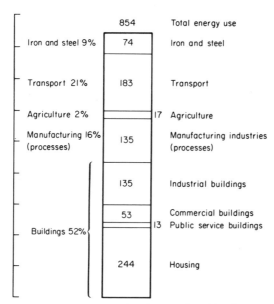

*OECD Europe includes 20 countries—see list given on the table of contents.
Notes: Total Delivered Energy Use does not include feedstocks or petroleum products consumed for non-energy uses.

Energy use in industrial buildings is estimated to be ½ of consumption in other industry, chemical and petrochemical industry sectors. British Gas estimate that 60% of industrial energy use goes to buildings. An estimate for energy use in agriculture or buildings has not been made.

Figure 14.11. Delivered energy demand in OECD Europe* by sector—1977.

oil shortages, or indeed to reduce oil demand. In 1979, a 5 per cent cut in oil consumption was called for in the United Kingdom. In transport, British Rail were alarmed and concerned about cutbacks in train services. Industry was equally anxious about reducing production.

Yet a 5 per cent cut overall is itself unlikely to produce permanent savings. It could mean purely a temporary inconvenience, possibly a reduction in industrial output and a lowering in material standards. This is where buildings offer a choice. For example, the equivalent of a 5 per cent national cut in oil consumption could be achieved by 30 per cent savings in all oil heated buildings. These savings are large enough to be permanent and not rejected as inconveniences to be abandoned once temporary shortages are gone. Oil could also be switched out of some buildings altogether to alleviate shortages in transport and industry, while permanent measures in those sectors have time to bite. Actually, some switching from oil in buildings has occurred already. At least 16 per cent and possibly as much as one-fifth of Britain's oil goes to buildings. The 1973 energy crisis resulted in a 10 per cent drop in fuel oil for central heating in all buildings types. But central heating oil consumption climbed back up to its 1973 level by 1977. However, in 1978, consumption fell by 5 per cent despite the coldest winter since 1973 and a slight rise in total oil consumption.

Energy options in buildings for the future

What then are the questions associated with adopting such a policy in order to keep our energy options open for industry and transport.

Many technical studies have been carried out to demonstrate the potential for and the actuality of energy savings in the building environment (*see References 9 and 10*). The potential and the reality is large indeed, and the evidence in favour of energy conservation investment is clear. However, the unfortunate fact is that until recently the price and availability of fuels has been such that only the very large users, the wealthy, the committed and the far sighted, have felt able to take the necessary decisions. The success of energy conservation to date has been by no means small, neither has it been overwhelming. It is only now, i.e. in the summer of 1980, that industrial organisations (for example) are finding that their fuel bills are equalling their profits, and that in their next financial year their fuel bill will be going up at a rate considerably higher that the rate of inflation. No doubt historic cost conventions have added to this time delay.

In the United Kingdom the majority of our existing buildings stock is owned by Government (in its various forms and manifestations); for example, 50 per cent of our domestic premises are owned by public authorities, two-thirds of our commercial buildings are rented by public authorities, and over 50 per cent of our industrial buildings are owned by state undertakings. The majority of new buildings are commissioned one way or another by Government, and it is said that in the United Kingdom over 90 per cent of the average architects design fees come from some form of State organisation. Perhaps in part for these reasons, there has been a tendency over the past decades for the building industry to be used as an economic mattress from which many things may be sprung. Such a situation has certainly resulted for whatever the reasons in an industry of warring barons, large and introspective, with an almost inevitable mismatch between availability of materials, labour, money and opportunities.

The energy used in the building sector is so large, that even a relative change in tariff structures between fuels, such as occurred in the United Kingdom in 1979, can rapidly produce a swing in demand such that a particular fuel supply 'runs out'. Such a situation tends to feed on itself, escalating very rapidly, so that the shortage of the particular fuel is soon matched by a shortage of the components necessary to use that fuel in our buildings.

If it is accepted that no one particular Government can control fuel prices for long, then can this option of interfuel substitution coupled with energy conservation in the building environment ever be realised in practice. Or is it inevitable that swings in fuel prices will occur and produce a swing in demand for the fuel itself followed by a shortage of the components necessary to use that fuel efficiently itself followed or proceeded by a shortage of the fuel itself etc.

The answer lies we believe in two factors, both of which lie at the economic/technical interface, and it is in effect these that we offer for perusal and comment to this IAEE Conference.

If the Government cannot control fuel prices, it can we would argue and indeed has been shown to be able to have major effect on the price margins between various fuels. This is of critical importance in the United Kingdom which is a multifuel economy. For it is in effect the margins between the fuels, and the anticipated future trends of those margins that determine individual (and corporate) decisions on particular fuel use. If the margins change too widely, and too quickly then there is a large movement of 'use' from one fuel to another. If the margins change slowly and over a longer time this fuel use change can be controlled. This one fact is supported or complicated by another namely the 'time-availability' for change. This latter may be explained best by an example, namely, if an industry has just committed itself to a new electric furnace and indeed installed one, it would take an enormous swing of marginal price to persuade that industry to change that furnace in the short term. If however the same industry were only considering the question of furnace change, then a much smaller change in price margin could cause them to take the same decision.

The argument is sometimes put another way. If one decides that we have a future then that future is concerned with electricity. The further one looks ahead the larger the role that electricity would play. Almost everything that we wish to do to conserve energy in our buildings increases our dependence on electricity. Therefore a planned process of energy conservation could be regarded as being concerned with the time scale at which we will move sectors of our buildings over to electrical usage. It would seem therefore that a better knowledge of the time scale potential for change of fuel use in our buildings is an important element in this process. This potential/ opportunity for change can be regarded as occurring at three scales, namely,

(i) Individual buildings
(ii) Urban areas
(iii) Regional areas.

Individual buildings. It is now common knowledge that whereas the fabric of a building lasts for a hundred years, the system of heating/cooling within a building lasts for a much shorter time, perhaps fifteen to twenty years.

We actually know a lot about the fabric age of our building stock, and also a great deal about how to improve that fabric. Improving the fabric does of course improve the potential for energy saving such that the return on investment tends to be longer than people have thought, as to fully realise the rate of return, the heating system itself also needs to be improved.

Whereas a long steady programme of improvement of our building fabrics is necessary to conserve energy, both because of the number of buildings that we have, and because of the problems associated with the building industry, this for the reasons given above is not enough.

It is important to know what the age is of the heating system when the fabric is being improved, or whether there is a system installed in the building at all. In terms of 'return' it is better if the fabric and the system are improved together as the potential for energy saving can be realised much more rapidly than if the fabric is altered first and the system altered ten years later. Clearly also the fabric must be improved first otherwise the improved system will be of the wrong size (e.g. the GLC problem). Now figures are becoming available (*see Reference 11*) of the numbers and types of heating systems in our buildings will occur and by 'using' the fuel margins to determine in which fuel direction they will occur. The problems are twofold. Firstly, the data is only beginning to appear, such that if this idea was felt to be valuable, it would be necessary to put considerably more time and effort into collecting this data. Secondly, the analysis and predictions necessary following from the accumulation of this data are not purely technical in their nature and the economic skills necessary to parallel the technical predictions are not those normally found in the building industry. An example of such data is that the domestic night storage heaters installed fifteen years ago which needed a lunch-time 'boost' are now nearing the end of their lives and are being thrown out. These systems are currently (because of the fuel price margins) being replaced by gas systems. Thus the lunch-time 'valley' in our electricity supply is re-appearing. Is this a desirable interfuel substitution? Should this electricity 'valley' be taken up by an increased gas load in our buildings, should new electrical tariffs be introduced to use up this peak or should the DeNorvic Pumped Scheme be used to 'take up' this 'valley'. A conservation opportunity with many options exists . . . the households be encouraged to go back to coal.

District mains. It has been said that some of our district electrical mains are ageing and up to capacity, that we are for example currently depending on copper put in the ground by a previous generation. It is also argued that similarly some of our gas district mains are coming up to capacity. Again marginal changes in fuel prices can either impose or relieve additional strains on these mains. Marginal changes in fuel price which result as the opportunity occurs in heating appliances being changed in individual houses will themselves in turn have a 'knock-on effect' on the mains. An interesting conundrum in which the real costs/benefits of, for example, district heating have their place. As once knowing the opportunities for change in the heating systems in any area, and knowing the capacity of the mains in that area, it is possible to present the economics of district heating in a very different light than at present.

Urban/Regional Scale.

The biggest opportunity for change occurs when either a local or the national authority decides to re-develop an area for example the inner city areas. Such re-development necessitates the taking of large numbers of energy/fuel decisions. Knowing the time scale for such developments (and even if all the lights are green it takes between five and seven years to get a building of any magnitude off the ground) it becomes possible to significantly alter the fuel mix in a particular location. For example the introduction of an MD tariff for commercial buildings in Eire 're-altered' electricity for industrial use elsewhere.

The purpose of this note has been to encourage those with skills not generally available in the building industry but in our view very necessary to it to consider the role energy conservation in buildings associated with a process of inter-fuel substitution may play in maintaining the flexibility of the fuels so necessary for industrial and transportation development. Such views if accepted of course require new multi-disciplinary and, we suspect, fairly expensive and long term studies to be carried out. The problem is however retractable and the rewards in our view worthwhile.

REFERENCES

1. 'Energy Balances of OECD Countries' 1975-1977. OECD, Paris, 1979 and 1960-1974.
2. 'Energy Consumption in the Soviet Bloc' Benedikt Korda, Vienna Institute (Wiener Instituts für Internationale Wirtschaftsvergleiche), 1977.
3. *'World Energy Supplies'*, J. Series, United Nations, New York, 1978.
4. 'United Nations Economic Commission for Europe' Energy/R.6/Add.2, 1979 Geneva.
5. 'British Gas Facts and Figures 1979/80' Edition, British Gas Corporation, London.
6. 'Digest of United Kingdom Energy Statistics', 1978, HMSO, London.
7. *'Rocznik Statystyczny 1979'* (The Polish National Yearbook) ROK XXXIX— Warsaw.
8. 'National Overall Energy Balances', Statistical Commission and Economic Commission for Europe, Conference of European Energy Statisticians, CES/AC.32/15, 10 October, 1977, Geneva.
9. 'Advisory Council on Energy Conservation Energy Paper 15' H.M.S.O. 1978.
10. 'Buildings—The Key to Energy Conservation' RIBA October, 1979.
11. 'Paying for Loft Insulation' National Consumer Council May, 1980.

PART VII

Aspects of New Energy Supply

Chapter 15

The Orinoco Petroleum Belt

Alirio Parra *

Often, the Orinoco Belt is mentioned together with such non-conventional sources of hydrocarbons as shale oil and tar sands. This has resulted in some misconceptions about the characteristics and feasibility of developing the oil from the Orinoco Belt. Therefore, I think that it would be useful for me to sketch some of what we now know about the size and nature of the Orinoco resource base and potential reserves, the cost of producing them, and the program that Venezuela already has underway for oil production from this region. The estimates are the latest available from the staff of Petroleos de Venezuela.

Over the next twenty years, we plan to raise production from the Orinoco Belt from its present very low level to at least one million barrels per day. This compares with Venezuela's present total output of about 2.1 million barrels per day. Over the years ahead, as the yield from Venezuela's older producing wells falls off, it is apparent that the production planned from the Orinoco Belt will assure that Venezuela will be able to sustain its present level of production, and even raise it to our goal of 2.8 million barrels per day of production potential. This will be true even if Venezuela's present exploration

*Director, Petroleos de Venezuela.

153

for offshore and other additional oil reserves does not yield the hoped-for results.

Our present plans for Orinoco Belt production are based on a conservative assessment of the fiscal needs of the country and of the demand for this type of oil. In fact—and more likely—the Orinoco Belt will have oil reserves which may be comparable in magnitude to the total of the present oil reserves of the rest of the world. This would suggest that even considerably higher levels of production would be eventually possible for centuries.

THE ORINOCO BELT

Probably the world's largest essentially untapped pool of oil resources, the Orinoco Belt, lies in South-Eastern Venezuela just north of the Orinoco River. It stretches about 600 kilometers from east to west, varies in width from 50 to 80 kilometers, and covers a total area that is larger than Ireland and Wales.

The oil-bearing sands of the Belt are from 600 feet to 7,000 feet deep, and their characteristics are very like those of the heavy oil fields of the Lake Maracaibo basin. The oil is indeed heavy. Although some of it has a higher gravity than 15 degrees API, and a considerable amount ranges between 10 degrees and 15 degrees, most of this oil is less than 10 degrees API.

Our present information is too limited to give a very confident estimate of the amount of oil in place in the Belt. A first estimate, made over a decade ago, was that the Belt held 700 billion barrels of oil. Information obtained since has led to upward revision of one trillion barrels, two trillion barrels, and even up to three trillion barrels. But it must be realized that, because of the physical characteristics of the reservoirs and the oil itself, these are not the kind of reserves that can be rapidly translated into production rates normally associated with the giant oilfields of the Middle East. Development will therefore be constrained to a much slower pace.

THE RECOVERY PROBLEM

The question of how recoverable they are is just as pertinent as the absolute size of the Orinoco oil resources. Since these deposits are so often compared with tar sands and shale oil, many people are surprised to learn that oil flows from wells in the Belt, and that no unusual technology is needed to produce this oil—even the heaviest. For many years, fields on the northern fringes of the Belt have been productive, and they are now putting out about 120,000 barrels per day.

There is more to the story than this, though. Because of the heaviness of the oil, standard primary production method will only get a very small percentage of these oils out of the ground. The most advanced secondary production techniques must be applied to make Orinoco Belt production practical and efficient. Fortunately, Venezuela has many years of experience with using such methods of steam stimulation, because they have been needed to produce the heavy oils of the Maracaibo Basin. About a third of Venezuela's present oil output is in heavy oils.

Without getting into the technology, I can summarize by saying that the techniques in common use around Lake Maracaibo often increase the rate of recovery of the heavy oils to more than 30 percent. We do not know yet whether results this good will be possible at comparable costs in the Orinoco Belt, but the results of testing so far are encouraging. Combining a bare minimum rate of recovery (10 percent) with the lowest estimate of oil in place (700 billion barrels) gives a low estimate of potential reserves that can be recovered of 70 billion barrels. Combining a 30 percent rate of recovery with the highest present estimate of oil in the ground produces a potential recoverable reserve figure of 900 billion barrels. For planning purposes, a figure of 500 billion barrels would not seem unreasonable. This is well over twice the proven reserves of Saudi Arabia.

FACTORS OF COST

The oil that we know is recoverable economically from the Orinoco Belt permits us to plan confidently for the production level that I mentioned earlier of one million barrels a day by the year 2000. There remain, however, some big unknowns that will affect the costs of recovery, and these factors in turn will influence the pace of exploitation of the Orinoco Belt beyond what is already planned.

For example, a two-stage process of secondary recovery is used very effectively in the Lake Maracaibo region, where a process known as steam soaking is followed up by one called steam driving. The first stage is highly efficient in Lake Maracaibo production of heavy oils because of particular geological factors present there. (This is known as compaction.) We do not yet know whether steam soaking will be as effective in the Orinoco Belt, although we have preliminary indications that it may be. If so, a relatively less intense use of the costly second stage, steam driving, will be needed to attain a recovery rate over 30 percent.

If conditions are unfavorable, as little as 8 percent of the oil in place would be brought to the surface through the low-cost steam soaking process, but when conditions are good, this is increased to a range of 15 percent to 24 percent recovery of the oil in place. Since a ton of steam can produce 25 barrels of oil with soaking, but only about three barrels with steam driving, you can see that the costs of production in the Orinoco Belt will be enormously affected by what we find out about the geology of the area. In terms of investment, the difference is a low-side estimate of $3,000 up to $10,000 to develop each one barrel a day of production potential.

The oil recovered from the Orinoco Belt, like Venezuelan heavy crudes, is high in sulfur, nickel and vanadium. These components must be removed before the heavy oils can be run through conventional refineries. This upgrading adds a cost to the use of heavy crude oils, but it is important to emphasize that acceptable technology is available, just as acceptable secondary recovery methods needed in production are at hand.

We do not yet have figures that have been sufficiently tested and examined to permit us to say with certainty what it will cost to produce a barrel of oil from the Orinoco Belt. We do know, however, that the costs are very competitive with those for shale oil and tar sands, and compare very favorably with costs for coal gasification and liquefaction.

ORGANIZATION FOR PRODUCTION

Venezuela's program for the exploitation of the Orinoco Belt is being carried out by my company, Petroleos de Venezuela. After some initial exploration of the Belt, the government just two and a half years ago gave Petroleos de Venezuela the responsibility for all future operations in the area. This, of course, is our most ambitious undertaking, and since given this responsibility, we have organized to carry it out.

We have established a new department to coordinate all activities in the Belt. In addition, we have given a major overall research responsibility to INTEVEP, the research and development arm of Petroleos de Venezuela, and have divided the Belt into four areas, and assigned one to each of our operating subsidiaries, LAGOVEN, MARAVEN, CORPOVEN and MENEVEN.

We have evaluated the results of exploration done by the Ministry of Energy and Mines before Petroleos de Venezuela was given the task, and last year we launched our own intensive exploration effort. By 1983, we expect to pull together the results of this varied effort, involving an investment of about $430 million. These results will provide the basis for firm development plans for the Orinoco Belt, beyond those we have already made.

Among other projects, the program that is now going on includes shooting some 16,000 kilometers of seismic lines, and drilling about 500 exploration wells and 1,100 evaluation wells. Among other important things, this work will determine the most promising area for concentrating initial development efforts.

Two of the projects are especially notable. LAGOVEN is carrying out one of them, under which 125,000 barrels a day of upgraded crude oil is to be produced by 1988 from an area in Monagas State. Under the second project, MENEVEN plans by the same year to be producing 75,000 barrels a day of conventional heavy crude oil for export, using existing production facilities in the south of the State of Anzoátegui. These two projects will enable us to assess the relative merits of upgrading the heavy crude oil ourselves in the field, as compared with exporting the heavy oils for upgrading abroad by customers. Different approaches to secondary recovery will also be tested carefully through a series of pilot projects, of which an important one will be producing the crude oil for the LAGOVEN upgrading project.

CONCLUSION

From my brief account, I think you will agree that the Orinoco Belt should not be classified with the nonconventional sources of oil with which it has been compared for many years. This was understandable in past years, when oil prices were below the production costs of oil from the Orinoco Belt as well as the cost of oil extracted from shale oil and tar sands. Cost considerations now favor the development of oil from the Orinoco Belt, and we expect to be producing 200,000 barrels a day by 1988, 600,000 barrels a day by 1995, and a million barrels a day by 2,000.

These goals are within the reach of present technology, and the highest estimated costs. But our present plans must be regarded as highly modest and conservative. The potential of the Orinoco Belt is immensely larger. Venezuela is not only assured of being able to maintain its desired production potential, it is justified in thinking in substantially larger terms. Rates of production of two or three million barrels of oil per day from the Belt seem entirely reasonable to consider by the turn of the century, and these rates could be sustained literally for centuries from the oil wealth of the Orinoco Belt.

Certainly, this is a resource with a significant part to play in the transition that is underway in the world economy from its present dependence on crude oil, to a future when energy may cost more, but also when more of it will come from renewable sources that are dependable for the long term.

Chapter 16

France's Nuclear Power Programme

A Ferrari*

The large French nuclear power programme enters quite naturally in the energy context of our country. In 1979 France used 190 MTOE but only produced 50 MTOE which is 25 per cent. Objective of the French government is to limit dependency to 55 per cent in 1990. Fossil fuel reserves in our country are extremely low, thus we shall direct our efforts on energy conservation, new energies and especially on nuclear energy. In 1979 nuclear electricity represented 4.5 per cent of our energy balance; it should take a 30 per cent share in 1990. Such nuclear choice is justified by economic considerations. Indeed the cost of a nuclear kWh, 13.5 centimes, with a plant starting operation in 1990 is much lower than the cost of a conventional thermal kWh, 25 centimes with coal, and 36 centimes with fuel. We built several gas-graphite reactors in the 1960s, then in 1969 we shifted to PWRs which are more economical. The first five units of this type of reactors built in France by Framatome with a Westinghouse licence were committed before 1973. Following the increase of oil prices and the growing scarcity of oil, the 1974 crisis led to accelerating the nuclear power programme. Since that time the French nuclear power programme has been developing quite

*Deputy Director for Programming and Planning CEA, France.

satisfactorily and it was not influenced by the slowing down or even the recession of the nuclear industry which spread out in most countries of the western world.

Today there are in France 18 installed units, 15 of them are in industrial service. They represent a total of 11 GWe net. Eight of these nuclear units are older (6 gas-graphite reactors, 1 gas-cooled heavy water reactor and 1 PWR), the other units are 9 PWR 900 and the Phénix breeder prototype.

32 units are presently on construction, 22 of them are PWR-900, 9 are PWR-1300, the last one being SuperPhénix the 1200 MWe fast breeder reactor. These units represent a total 32.6 GWe net and beginning of industrial service shall extend from 1980 to 1986. This equals an average increase of the nuclear electricity of more than 5 GWe per year. In 1986 installed power will thus reach 44 GWe.

A further development of the French nuclear programme in the short term considers beginning of construction of 6 units until the end of 1981: 2 PWR-900 and 4 PWR-1300, this is a total of 7 GWe net; the industrial service of which should take place between 1986 and 1988. Five units will be launched after 1982: 1 PWR 900 and 4 PWR-1300, or a total of 6 GWe net. Beginnning of industrial service is expected in 1987 and 1988.

After 1988, 3 to 5 PWR-1300 shall start up each year, and the first fast breeder following SuperPhénix shall start operation around the end of the decade. Once this programme has been carried out, installed power in France will rise from 13.6 GWe in 1980 to 44 GWe in 1985, and to 62 GWe at the end of 1990. For the year 2000 the forecasted nuclear power ranges from 85 to 110 GWe.

The considerable and rapid development of nuclear energy requires a great technical, industrial and financial effort. The cost of investment of a nuclear plant reaches around 3,800 F/installed KWh; this is 4 billion FF for a 1,300 MWe PWR. However such investments are more than justified by an increased energy independence and by the lowered production cost of electricity. Nuclear boilers are built by Framatome, which has already produced 32 reactor vessels, its production capacity is 8 vessels/year. Furthermore, together with CEA and EDF, Framatome is carrying out a large R and D programme, more particularly in the fields of reliability and nuclear safety. Thanks to these efforts it was possible to adjust the PWRs to the French operating conditions and to meet the particularly strict safety standards which the French national regulations require. In addition to its building of the French reactors Framatome is also in charge of exporting nuclear devices, from the nuclear vessel to the turn-key power plant. France's seeking for its energy independence implies not only to build and operate nuclear power plants in sufficient number, but also to control all the steps of the nuclear fuel cycle, from the uranium production to the reprocessing of irradiated fuels. This major sector of the nuclear industry is controlled by Cogema, a 100 per cent subsidiary of CEA, but a number of large private groups also have interests in this sector.

Proven resources of natural uranium in France come to 120,000 tonnes, or 25 per cent of the total world proved resources. Should these resources be used in light water reactors only, they would represent 1 billion tonnes of oil equivalent; with the breeders their energy potential will be multiplied by about 80. Today ore concentration is performed in 6 plants and 3 new facilities shall start operation in the early 1980s.

In 1979 the national production was slightly above 2,000 tonnes, from 1980 onwards it should overstep 3,000 t/year. The production covers 50 per cent of the domestic demand at the present time; at the end of the decade the percentage will still be close to 35 per cent. Participation of French companies in uranium mining abroad and purchases on the international market allow our supplies to be made up.

Enrichment services are one of the main concerns of the utilities because it is a preliminary condition for their fuel supply. France decided in favour of the gaseous diffusion process because of its technical and economic advantages. Following the start up and satisfactory operation of the Pierrelatte plant, construction of a large plant was decided at the end of 1973 and the multinational company Eurodif was created. Owing to the projected capacity: 10-8 MSWUs a year, and to the extent of the industrial and financial effort required, international collaboration was imperative.

The undertaking proved to be a success when enriched uranium was produced for the first time at Tricastin in 1979; full capacity will be reached in 1982; the whole project was achieved within the appointed time and with the initially determined budget, a total of FF9.5 billion (economic conditions of 1974).

Up to now we had to resort to enrichment contracts with the United-States and the USSR. The French part in Eurodif will supply our enriched uranium needs until the late 1980s. But as early as the end of the decade, or at the very beginning of the subsequent one, we will need a new plant to meet the increasing needs in France and in the world. A project is presently under review for the construction of a second gaseous diffusion plant in Europe or somewhere else, and development of the chemical process is actively carried on. This process proves particularly interesting for units with a small capacity (less than 1 MSWUs/year).

The reprocessing of irradiated fuels aims at separating products which can be used anew from the radioactive wastes contained in the fuel elements, which appears to us a necessary step for safe waste disposal. In addition this is a preliminary for the development of breeders which are part of the French programme. Since 1976 France dispenses at La Hague of a prototype workshop (high activity oxide). Its processing capacity is at the present time 200 to 250 tonnes fuel/year. The industrial programme on hand, under Cogema's responsibility and with CEA support, is centered around the construction of two plants, each having a capacity of 800 t/year, UP2-800 and UP.3.A, industrial operation of the first plant shall start in 1985. A large R and D programme is being carried out by CEA as a support for designing and building the reprocessing plants. The national effort is complemented by actions in collaboration with foreign partners, more particularly with Great Britain which follows nearly the same trends in the field of reprocessing. Joint negotiations have been initiated in 1978 by the French-British industry with the Germany industry.

As to the disposal of high radioactive wastes produced by reprocessing, French plants will be fitted out with vitrification workshops; the prototype vitrification workshop which has been operating at Marcoule since 1978 gives every cause for satisfaction.

With the breeder reactors quite considerable savings of national uranium will be possible. Breeders make a much better use of uranium than do light water reactors. Since the beginning of the 1950s studies are being carried out in France in this field. In 1961 construction of the experimental reactor Rapsodie began at Cadarache. The reactor became critical in 1967, one year later a 250 MWe reactor, Phénix, was launched. It began industrial service during the first half of 1974. Since that time Phénix has generated more than 7 billion kWh. Phénix gave the possibility to assess and to solve the specific problems posed by the French system of fast breeder reactors (integrated pool). Construction of a 1,200 MWe reactor, SuperPhénix was undertaken to corroborate on an industrial scale the success of Phénix. SuperPhénix is now on construction at Creys-Malville since 1977. Beginning of industrial operation shall take place in the winter of 1983-84. SuperPhénix is a prototype plant; cost price of the kWh is higher than with standard PWR; however it will remain close to the price of one kWh produced by a conventional thermal coal plant.

The next step in developing this type of reactor could be, depending from the technical and economic optimization studies on hand, the launching of a new programme with a small number of reactors and the corresponding fuel cycle facilities. Within this frame EDF could consider, prior to 1986, the construction of a first pair of 1,500 MWe reactors, in continuity with Phénix. The corresponding facilities of the fuel cycle (fuel manufacturing and reprocessing) would be undertaken simultaneously. The industrial structure which shall permit to ensure the development of the French fast breeders does exist now. It is composed, in addition to the Novatome Company established in 1976 which is in charge of engineering and construction, of the French System Company for Advanced Reactors (SYFRA, whose shareholders are CEA and Novatome), which is responsible for collecting knowledge and know-how of its shareholders to constitute the Licence Dossier. The Dossier will be then negotiated by SERENA (European Company for the Promotion of Sodium Cooled Fast Breeder Reactor Systems).[1]

Although it is a leader in this field, France has clearly announced that it did not want to continue individually the development of fast breeder reactors, these reactors are made absolutely necessary for a certain number of countries by the current energy situation. While protecting the results of years of efforts and the place of our country in this field, CEA, within the framework of the Government guidelines, has applied itself to develop contacts and agreements with the various national and international partners; more particularly with the Federal Republic of Germany with which France signed cooperation agreements as early as 1977. The agreements were concluded between industrial partners and research organizations in both countries. The industrial agreements between Novatome in France, NIRA[2] in Italy and INB[3] which is composed of industrialists of the Federal Republic of Germany, Belgium and the Netherlands bear evidence of the special links between the European countries; the negotiations undertaken by SERENA should result in industrial and commercial development with new partners through licence agreements.

As regards Research and Development, the agreements range from the comparison of general information on the reactor type up to the wide exchange of knowledge and coordination of programmes. We must add to the agencies of the abovementioned countries, those of the UK (UKAEA), USA (Department of Energy), Japan (Power Reactor and Nuclear Fuel Development Coorporation) and agencies of the USSR.

The French nuclear power programme, probably the most ambitious one today in the Western world is an essential element of the energy policy in our country. It complies with the requirements of security and economic efficiency. The programme is backed by a coherent R&D effort and by an effort of industrial development for the construction of the reactors as also for all the other steps of the fuel cycle.

NOTES

1. SERENA shareholders are KVG (a company created by German firms Interatom, KfK and Belgian and Dutch Parners) and the French company SYFRA.
2. NIRA: Nucleari Italiana Reattori Avanzati.
3. INB: Internationale Natrium Brutreacktor bau Gesellschaft.

Chapter 17

The Potential of using Biomass for Energy in the United States

*Wallace E. Tyner**

Today, oil and natural gas constitute three-fourths of US energy consumption (oil—one half, and gas—one fourth). But it is pretty clear that oil production in the United States has peaked and is on the decline. Today the United States is importing about one half of its oil needs and imports amount to about 3 billion barrels per year or $80 billion per year at $30 per barrel. Because of the balance of payments burden and more importantly, because of the perceived threat to national security posed by the high level of imports, the United States is seeking means of reducing total oil imports. In the next energy transition, most of the discussion centers around displacing oil and natural gas and reducing the rate of growth in energy consumption.

A number of options to accomplish this transition are being considered including the following:

(i) energy conservation
(ii) increasing the domestic oil supply
(iii) changing energy consumption from liquid to solid fuels such as coal and nuclear

*Department of Agricultural Economics, Purdue University.

(iv) converting solid fuels such as coal and shale to liquids

(v) increasing use of renewable energy sources such as solar and biomass

Due to space limitations, we cannot discuss each of these alternatives. We should note, however, that all of these options are aimed at reducing United States' dependence on foreign sources of liquid energy either by reducing total energy consumption or by converting from liquid to solid energy sources. It does appear that the energy strategy to be followed by the United States will include a combination of all these options.

For this paper we will concentrate on a portion of the last strategy—production of energy from biomass. In particular, we will concentrate on the potential of producing energy from agriculture. The agricultural energy potential includes both food/feed crops and cellulosic materials such as forage crops and crop residues. In this paper we will emphasize the interface between production of food/feed and energy and the policy choices important to this issue.

BIOMASS RESOURCE BASE

The biomass resource base includes a wide range of materials such as grains, sugar crops, food wastes, animal wastes, crop residues, wood, forest residues, and municipal solid waste. Several studies have attempted to assess the potential energy from each of these categories of biomass. The results from three of these studies are summarized in Table 17.1. All of the studies did not examine all resources. Many of the results exhibit a broad range of energy potential reflecting the inherent uncertainty in the estimates. The total potential for biomass energy in the United States in the next two decades is likely to range between seven and sixteen quads per year which is 9 to 20 percent of 1979 US energy consumption.[1] The largest single biomass resource is wood which may provide five to ten quads by 1995. The next largest resource is forage crops which could provide one to four quads. Municipal solid waste could provide 0.2 to 5 quads per year. The other categories produce one quad or less each. It is important to note that the grains category could yield .3 to .7 quads by 1995 which is 3.6 to 8.4 billion gallons of ethanol. That amounts to 3.2 to 7.6 percent of 1979 gasoline consumption in the United States. The grains component of biomass energy is about 4 percent of the total biomass energy potential for the United States. For

Table 17.1. Potential Biomass Energy in the United States (Quads[a]).

Resource	OTA	DOE	Tyner & Bottum	Author's Judgement[b]
1) Grains[c]	0-1	.25	.6-.8	.3-.7
2) Crop residues	.8-1.2	—	.6-.9	.6-1.0
3) Wood	5-10	—	–	5-10
4) Forage crops	.5-8.0	—	1.0-2.3	1-4
5) Food wastes	⟨.1	⟨.1	–	0-.1
6) Animal wastes	.1-.3	—	–	.1-.3
7) Municipal solid waste	–	.1	–	0.2-5[d]
Total				7-16

Sources: (1) OTA—Office of Technology Assessment,*Energy from Biological Processes,* Volume I. (Washington, D.C.: Office of Technology Assessment, US Congress, May 1980.) (2) DOE—US Department of Energy, *The Report of the Alcohol Fuels Policy Review.* (Washington, D'C.: Government Printing Office, June 1979.) (3) Wallace E. Tyner and J. Carroll Bottum, *Agricultural Energy Production: Economic and Policy Issues,* Station Bulletin No. 240. (W. Lafayette, Indiana: Agricultural Experiment Station, Purdue University, Sept. 1979.)

[a]A quad is one quadrillion BTUs. Current US energy consumption is about 78 quads per year. Note that the energy forms are not all comparable in terms of point of measurement. For example, the grain numbers are measured as alcohol produced whereas the wood is measured as direct combustion input, gas produced, and alcohol produced.

[b]The range represents this author's current judgement of the most likely range of biomass energy obtainable in the United States by 1995-2000. The total range is not the sum of the individual ranges because some of the categories represent use of the same land resource.

[c]The grains would most likely be used for ethanol production. One quad of ethanol is about 12 billion gallons.

[d]This estimate was derived from information in James R. Greco, "Energy Recovery from Municipal Wastes" in *Fuels and Energy from Renewable Resources,* edited by D.A. Tillman, et al., New York and London: Academic Press, 1977.

ethanol alone, the grains component may be as much as 25 to 30 percent of total ethanol production.[2] The point is that grains or food and feed crops generally are expected to supply only a small portion of the total biomass energy.[3] The remainder of the biomass resources do not directly compete in food/feed markets.[4] So the food versus fuel issue is relevant for only a small portion of the total agricultural and biomass resource base. It is important to keep this relationship in mind as we examine biomass energy potential and the food/feed/fuel question. To the extent that conversion technologies can be developed that efficiently convert solid biomass resources into liquid fuels, the lignocellulosic feedstocks (wood, forage crops, and crop residues) will be the major source of liquid fuels from biomass.

LIGNOCELLULOSIC BIOMASS RESOURCES

Space does not permit a detailed explanation of the resource base estimates for lignocellulosic resources. Municipal solid wastes will not be discussed further in this paper because the major emphasis is on biomass resources directly related to agriculture. We will, however, provide a brief description of the estimates for wood, forage crops, and crop residues—the major resources in this group.

Wood

Wood is the largest biomass energy resource in the United States. Wood resources include both forest residues and conventional wood harvesting. Most of the energy resource is expected to come from conventional harvesting, and much of it will be used in the forest products industry. Today that industry consumes about 2.7 quads total energy, of which about 1.2 quads is wood energy. Assuming the forest products industry doubles in size by 2000 as projected, it could be consuming four to five quads of energy per year, most of which would be wood. Wood can also be used to substitute for fuel oil in industrial or utility boilers or for home heating. In addition, wood can be gasified or converted to methanol for fluid energy. Achievement of the 5 to 10 quad level of energy from wood by 1995-2000 requires intensification of forest management. If a long term intensive forest management approach is not taken, existing forest resources could be depleted and significant environmental degradation occur.

Forage crops

Forage crop production increases were estimated for the US states of Iowa and Minnesota and all other states east of the Mississippi River.[5] These areas were considered to have adequate rainfall to support increased herbage production. A total of about 102 million acres in the eastern United States potentially could produce greater quantities of herbage through more intensive management. Excess production does not exist on these acres at the present time. Production above that needed for livestock feed can be obtained only to the extent it is profitable for farmers to fertilize and intensively manage these acres. The application of small amounts of fertilizer and changing harvest schedules could easily result in the production of an additional one to two tons per acre. On many pasture and hayland acres even higher production increases may be achieved. Through seeding of more productive species, higher levels of fertilization, and more frequent harvesting, an additional one to two dry tons per acre might be achieved. This author's best estimate is that one to four quads of energy could be obtained from forage crops by 1995-2000 without diverting any supplies away from livestock feed or other uses.

Crop residues

The total amount of crop residue produced in the United States each year is about 400 million dry tons.[6] To estimate usable crop residues, this base number was adjusted to allow for three factors: 1) maintenance of soil quality (prevention of soil erosion), 2) residue losses in harvesting, and 3) storage and transportation losses. The amount of residue needed to be left on the soil for soil conservation purposes depends on the soil texture and climate. The authors of the Purdue Study estimated the amount of crop residue which could be removed safely for each crop and US land resource area using original data developed by Larson *et al.*[7] In addition to deduction of residues for soil erosion control, any residue which could not be collected with current harvesting machinery also was deducted for each crop. Finally, storage and transportation losses were deducted to arrive at total usable crop residues. Total usable residues amounted to about 80 million dry tons per year or only about 20 percent of the total annual residue production. Of course, all of this could not be collected and used economically. The range of .6 to 1.0 quads reflects the uncertainty in the extent to which the crop residues could be economically collected and transported to a central facility or used on farm.

FOOD AND FEED CROPS

Grains represent the vast majority of US potential for energy production from food and feed crops. For that reason, sugar crops are not dealt with at length in this paper. It is clear that the United States can produce sufficient food and feed grains for domestic consumption now and in the foreseeable future. The only relevant question regards US capacity to meet both domestic and export demands as well as produce energy. To get a clearer picture of what actually happens at present to grain resources, it is instructive to examine the total disposition of US grain production.

The United States and world grain markets

Despite the fact that large quantities of agricultural commodities enter international trade, the fact is that most food and feed crops are consumed in the region in which they are produced. Less than 15 percent of the grain produced worldwide enters international trade.[8] A much larger fraction of US production enters international trade. The United States and other grain exporters become the residual supplier for some countries in years when their domestic production is lower than average. For other countries, US crops are imported each year as a normal part of their food and feed requirements.

The United States has become increasingly important as a source of food and fiber as world consumption has increased due to population and economic growth. The United States produces about 20 percent of total world grain production and consumes about 12 percent of the total. While the United States does not feed the world in the sense of producing a major share of the world's total grain output, US agricultural exports do play a critical role in world markets. They are especially important as a source of foreign exchange, a source of income for US farmers, and an important source of employment in the US economy.

The United States is the world's largest exporter of agricultural products. The United States share of total world agricultural exports has increased from 12 percent in the early 1950s to over 17 percent in recent years. About half the grain and soybeans traded in international markets come from the United States. In recent years about three-fourths of the total value of US agricultural exports has come from wheat, corn, and soybeans.

Wheat

Of the food grains wheat and rice, wheat is much more important in US production and trade.[9] About 57 percent of US wheat is exported, about 8 percent used for feed domestically, and about 35 percent used for food and seed domestically. In the export category, about 89 percent goes into commercial sales and 11 percent is used for all government concessionary programs. Europe accounts for about 14 percent of US wheat exports; Japan, 11 percent; USSR,

12 percent; and developing countries, 47 percent (concessionary plus commercial). Hence, about 27 percent of US wheat production ends up in developing countries. Since wheat is used directly for food and because a much larger fraction of US wheat production ends up in developing countries, the food–energy trade-off is very different for wheat and the feed grains.

Corn

Of the feed grains, corn is clearly the most important constituting about 86 percent of total US feed grain production. Currently, about 61 percent of US corn is used domestically for animal feed, 31 percent is exported, and 8 percent used for food, seed, and alcohol production. In the feed category, about 23 percent is for beef cattle, 17 percent for poultry, 34 percent for hogs, 17 percent for dairy animals, with the remainder being used for other animals. In the export category, about 4 percent goes for concessionary programs of all kinds including P.L. 480, donations, long-term credit sales, and aid. Most of the corn exports are commercial sales. About 46 percent of the exports are to European countries, 17 percent to Japan, and about 18 percent to the USSR. The total exports to developing countries (commercial plus concessionary sales) amount to about 13 percent of all exports or about 4 percent of total disappearance.

Corn is the grain most often discussed as a feedstock for alcohol production in the United States. At current market prices corn is a cheaper feedstock than wheat, sugarcane, sugar beets, or molasses. Since corn is a feed grain and is not consumed directly by humans in large amounts, a more correct characterization of the food/fuel issue would be the food/feed/fuel issue. The use of large amounts of corn for ethanol production would tend to increase corn prices and lead to reduced use of corn for animal feed. The corn price increase would lead to higher meat, dairy, and poultry product prices which would, in turn, lead to reduced consumption of these products. The initial impact probably would be a shift in the consumption pattern among animal products, e.g., a shift from beef to poultry, a more efficient converter of corn. In other words, the major effect of corn use in ethanol production would be changes in the diet in countries where animal products make up a significant portion of the food intake. From this perspective, the major impact of a large US grain alcohol program would be on people (throughout the world) who consume animal products.

Alcohol production potential

The discussion of alcohol production potential will focus on the potential from grains because that is the source closest to the food/fuel issue and the fuel attracting most of the political and public interest. The next major question concerns the energy production which could be achieved on the new acreage. To answer this question we must understand the relationship between alcohol production and the protein by-products of the process. One bushel of corn (56 pounds, 25.4 kilograms) yields 2.57 gallons (9.7 liters) of 200 proof anhydrous ethanol plus 17.35 pounds (7.9 kilograms) of distillers dried grains (DDG). About one pound of distillers grain is produced for every three pounds of corn used. Distillers grain is about 27 percent protein and 12 percent fiber.[10] It is useful as a protein supplement for beef cattle and dairy cattle. In that role, it substitutes for soybean meal and other high protein animal feeds. About 1.63 pounds of distillers grain contain the protein in one pound of soybean meal.[11] Potentially distillers grain could replace much of the soybean meal in cattle diets. To the extent this substitution occurred, soybean acreage would decrease and corn acreage increase.

If the wet milling process or modified dry milling process were used to make alcohol from corn, the protein by-products would be gluten meal and gluten feed instead of distillers grains. Gluten meal with 60 percent protein is much higher in protein than either distillers grain or soybean meal. Gluten meal is much lower in fiber than distillers grain and therefore suitable for poultry and other animal feed diets.[12] Corn oil is another valuable by-product with these processes.

Some have argued that because these high protein by-products are created when alcohol is made from corn, using corn for fuel causes no conflict with food use. The argument is that protein is the limiting food ingredient and that the alcohol production only uses the starch and saves the protein. The latter point is correct—the corn starch is converted to sugar and fermented to alcohol. However, it is incorrect to say that protein is the limiting ingredient in food or animal feed. As discussed above, most of the US corn goes into animal feed, either domestically or after export. The primary function of corn in animal diets is to provide energy which comes from the corn starch. So it is erroneous to believe that there is no opportunity lost of removing the starch for alcohol production. In fact, it is the starch which gives corn its high value as an animal feed.

We can now use the information on protein substitution to obtain an idea of the amount of new grain acreage required for alternative

Table 17.2. Corn Acreage needed for Alcohol Production.

Ethanol production (bil. gal./) year)	Corn required (mil. bu.)	Soybeans displaced (mil. bu.)	Soybean acreage reduction (mil. ac.)	Corn produced on soybean acres (mil. bu.)	Additional corn required (mil. bu.)	New corn acreage (mil. ac.)
1	389	86	2.88	259	130	2.0
2	778	173	5.75	518	261	4.0
3	1167	259	8.63	776	391	6.0
4	1556	345	11.50	1035	521	8.0
5	1946	431	14.38	1294	651	10.0
10	3891	863	28.76	2588	1303	20.0

ASSUMPTIONS:
1) Corn average yield per acre on existing corn/soybean land = 90 bushels.
2) Soybean average yield per acre on existing corn/soybean land = 30 bushels.
3) Average corn yield on new corn acreage = 65 bushels.
4) DDG yield = 6.75 pounds per gallon of ethanol or 17.35 pounds per bushel of corn.
5) Alcohol yield = 2.57 gallons per bushel of corn.
6) DDG has 27 percent protein and soybean meal has 44 percent protein, and the substitution ratio is 1.63 pounds of DDG for 1 pound of soybean meal.
7) Soybean meal is 80 percent by weight of the soybeans (48 of 60 pounds per bushel).
8) The new corn acreage number represents the acreage addition required assuming the existing quantity demanded for protein and starch is satisfied in addition to the alcohol demand for starch.

CALCULATION PROCEDURE:
1) column 2 = (col. 1)/2.57
2) column 3 = (col. 2) × (17.35/(48 × 1.63) = (col. 2) × (.2217)
3) column 4 = (col. 3)/30
4) column 5 = (col. 4) × 90
5) column 6 = (col. 2) − (col. 5)
6) column 7 = (col. 6)/65

levels of alcohol production (Table 17.1). Column 1 of Table 17.2 contains the alcohol production level, column 2 the corn required for that production, and column 3 the soybeans which could be displaced by the protein by-product (DDG). Column 4 contains the soybean acres displaced, column 5 the potential corn production on the displaced soybean acreage, and column 6 the increase in corn production required to achieve the stipulated alcohol production level. Column 7 contains the increase in corn acreage required to achieve the alcohol production. Corn and soybean yields were held at their 1977 levels of 90 and 30 bushels per acre respectively. Corn yields in 1978 and 1979 exceeded 100 bushels per acre. The yield of corn on new acreage was estimated at 65 bushels per acre on the assumption that land not now in production would be of lower quality on average than existing corn land.[13] The results of these computations indicate that each billion gallons of alcohol produced from corn would require two million acres of new land in corn production. In 1980, about 80 million acres will be planted to corn and 70 million acres to soybeans in the United States. In 1978, there were about 13 million acres in corn and wheat land diversion programs (set-aside) in the United States, of which 4.6 million acres was corn land. These figures indicate that 2 billion gallons of alcohol could be produced with a net addition of 4 million acres of corn land, which is less than the 1978 set-aside.

In previous work, Meekhof, Tyner, and Holland estimated the impact on corn and soybean markets of levels of alcohol production ranging from 1 to 4 billion gallons using a stochastic simulation model.[14] In that work, levels of 1 to 2 billion gallons had relatively little impact on corn and soybean prices and quantities. However, levels of 3–4 billion gallons tended to increase corn prices substantially and to increase price instability in both corn and soybeans.

These results all tend to indicate that at relatively low levels of alcohol production (1–2 billion gallons), the corn price increases and required land supply increases would be relatively small. At higher levels of alcohol production, the impacts would be more serious. The econometric model results probably overstate the price impacts because long run adjustments would tend to change production patterns more than indicated by the model. Nonetheless, US policy makers probably should establish current policy for alcohol production with the intention of reviewing the impacts and prospects again once production reaches 1–2 billion gallons per years.

Grain alcohol economics

The economics of producing alcohol from corn depend on the price of corn, by-product prices, and the size of the plant. Table 17.3 summarizes the economics of corn alcohol for a 50 million gallon (189.3 million liter) plant. At $2.50 per bushel corn ($98.42 per metric ton), alcohol would cost about $1.21 per gallon (32 cents per liter) in first quarter 1980 dollars. This cost increases to about $1.40 per gallon ($.37 per liter) for a 20 million gallon (75.7 million liter) per year plant and to about $1.55 per gallon ($.40 per liter for a 10 million gallon (37.9 million liter plant).[15]

To put these numbers in perspective, the US refinery gate unleaded gasoline price in first quarter 1980 was about $.95 per gallon ($.25 per liter) which implies an average crude oil acquisition price of $25 per barrel. We can approximate the conversion of crude oil

Table 17.3. Cost of producing Alcohol from Corn.

Corn price ($/bu.)	Corn cost[a] ($/gal.)	Fixed costs[b] ($/gal.)	Operating costs[c] ($/gal.)	Gross Total cost ($/gal.)	By-product credits[d] ($/gal.)	Minimum alcohol selling price[e] ($/gal.)
1.50	.58	.31	.31			
1.50	.58	.31	.31	1.20	.27	.93
2.00	.78	.31	31	1.40	.33	1.07
2.50	.97	.31	.31	1.59	.38	1.21
3.00	1.17	.31	.31	1.79	.44	1.35
3.50	1.36	.31	.31	1.98	.49	1.49
4.00	1.56	.31	.31	2.18	.55	1.63

Source: The raw data for compiling this table was obtained from *Grain Motor Fuel Alcohol-Technical and Economic Assessment Study* prepared for the US Department of Energy by Raphael Katzen Associates (June 1979).

Note: The data was converted to first quarter 1980 costs. Plant size is 50 million gallons per year. The plant uses coal for processing energy.

[a]This conversion assumes 2.57 gallons of alcohol is produced from one bushel of corn.

[b]The fixed costs include amortization of the investment cost over 15 years at 15 percent rate of return plus license fees, maintenance, tax, and insurance. The capital cost including working capital is $73.14 million.

[c]Operating costs include raw materials other than corn, such as energy, labor, overhead, freight and miscellaneous expenses.

[d]The by-products are distillers grains and ammonium sulfate. Distillers grains are valued at $110 per dry tin and contribute 37 of the 38 cents by-product credit at $2.50 corn. The distillers grain by-product credit in this table is calculated assuming that protein prices change along with corn prices. We assumed that three-fourths of the proportional change in corn prices occurs in DDG prices. For example, the change in corn price from $2.50 to $3.50 represents a 40 percent increase in corn prices, and we assumed that DDG price in that circumstance would be $143, which is 30 percent higher than the $110 base price.

The two prices would tend to move in the same direction because corn and soybeans (a protein source) are grown in the same areas on the same type of land. However, the exact relationship between the two prices is not known, and this assumption represents our judgment as to what the relationship might be.

[e]This price includes profit for the producer (in the capital recovery component of fixed

acquisition price to refinery gate gasoline price by multiplying the crude price by 1.6 and dividing by 42 gallons per barrel (a factor equal to .038).[16] With an average crude oil acquisition price of $35 per barrel, the gasoline refinery price would be about $1.33 per gallon ($.35 per liter) which is higher than the alcohol price for a 50 million gallon plant. Hence, even if alcohol from corn is not quite economic today, it will be within eighteen months.

Also, ethanol has higher octane than gasoline and this octane enhances the value of alcohol over gasoline in some circumstances. The refinery output of unleaded gasoline can be produced at a lower octane when the gasoline is to be blended with alcohol, resulting in substantial energy and economic savings in refining. With this refinery savings, the Office of Technology Assessment estimates that alcohol produced from corn is economic today in the United States without any subsidy.[17]

CURRENT POLICY SITUATION

The nature of the food/feed/fuel issue is quite different depending on whether alcohol or gasoline has a higher market value. At present, alcohol is more expensive than gasoline in the United States and must be subsidized to be competitive. In the future petroleum prices may rise faster than grain prices giving alcohol the competitive advantage over gasoline. In the first situation, the United States might be subsidizing the diversion of food/feed grains from food/feed uses to fuel use. In other words, the US government would be making a conscious policy decision to utilize a portion of its food/feed for fuel. Depending on the level of fuel utilization, this policy choice could mean the diversion of food/feed from commercial or concessionary export markets. On the other hand, if alcohol becomes less expensive than gasoline, grains would be diverted to fuel use without any policy intervention. In fact, the US government could find itself in the position of choosing whether or not to prevent grains from being used for alcohol to maintain concessionary or commercial grain exports or to minimize grain price increases for animal feed. The point is that any moral question regarding food/feed versus fuel is cast in a different context if the government is actively subsidizing food/feed use for fuel. If the grain for fuel allocation is market determined, it is no different in principle from use of grains for beef cattle, corn sweeteners, or industrial products when people are hungry or malnourished elsewhere in the world. Thus the relative prices of corn and oil help to determine the context for analysis of policy issues.

From the background information provided above, it is clear that any moral questions associated with the use of grains for alcohol fuel are not so clear cut as they may appear on the surface. Much of the US grain production is used for feed and not directly for food. Less than 4 percent of US corn production is consumed in developing nations as food or feed. Use of feed grains for alcohol involves trade-offs in the price and quality of meat, poultry, and dairy products, but not directly the issue of food for starving humans, at least at low to moderate levels of alcohol production.

The examination of major policy issues which follows should be viewed from this perspective. Current US policy regarding food aid, farm program management, and farm exports will be reevaluated in light of the potential for producing alcohol from grains. These policy areas plus the general topic of policies to ease the energy transition in the United States are discussed in this section.

Export policy

US grain exports to most of the grain importing nations have been relatively stable and rising through time. Much of the variance in US grain exports has been due to fluctuating demands of the Soviet Union. The USSR is becoming an increasingly larger purchaser of US crops, but the Soviet demands are quite erratic. Johnson has estimated that 80 percent of the variability in world wheat imports between 1963 and 1974 was due to variations in the wheat imports of the USSR.[18] If, in the future, the United States continues to supply the USSR with grains as needed, its grain exports will continue to be variable. Much of the Soviet grain is produced in relatively high hazard crop areas. Therefore, their grain needs vary widely from year to year. A major question in US agricultural export policy is should it continue to depend on such a highly variable market for its crops. This question would have arisen even without the recent Russian grain embargo. As the economics of grain alcohol production improves, the trade-off between a stable domestic market (alcohol) for a portion of its grains and a highly variable export market to the USSR will become more obvious.

Food aid policy

Over the last twenty years, a consensus has emerged that direct food aid has both harmful and beneficial effects. In times of famine or general food shortages, food aid can be quite helpful in saving human lives and improving nutrition. Long-term food aid, however, may create a dependence on the donor nation and tends to reduce food prices in the recipient nation. Reduced food prices diminish the incentives for farmers to increase domestic production. So long-term food aid can do more harm than good. The developed nations, including the United States, are now following a general policy of providing food aid in times of an emergency but not on a continuing basis—to help developing nations develop their own food production capabilities. This policy seems likely to continue. It does not seem likely that the United States will send great quantities of food to the food deficit countries on a continuing basis in future years. Of course, this does not mean that the United States will fail to supply food aid in emergency or famine situations. To the contrary, US food aid policy is to provide aid in emergency conditions, but not to build up a long-term dependence on US food aid.[19]

Farm program policy

In general, every aspect of farm policy will need to be reconsidered as energy production is added to the list of farm policy objectives. The land diversion and set-aside programs will need to be revised to encompass demand for grains for energy as well as for food/feed/fiber. Alcohol fuels legislation introduced, pending, or currently in effect can be classified in three broad categories: 1) policies to stimulate and stabilize the supply of agricultural raw materials for use as ethanol feedstocks, 2) policies to subsidize the conversion process, and 3) policies to subsidize end-product consumption.

Raw materials policies in the past have been designed to support farm incomes, stabilize commodity prices, promote farm exports, and protect consumers from excessively high food costs. The programs have included price supports and acreage set-aside provisions. They also have encompassed grain reserve policy. In all likelihood, a grain reserve for alcohol production will be needed in addition to the existing food/feed reserve. The reserve could be added to the existing reserve programs or a new targeted energy reserve to be held by alcohol producers could be created. One feature of the targeted energy reserve is that the grain clearly would be separated from existing food/feed markets and reserves, and there would be less tendency to draw on it for food/feed uses.

Other raw materials policies could include targeting a portion of the crop production for energy, or establishing commodity programs for cellulosic crops. Revisions in the acreage withdrawal (set-aside) program no doubt will be undertaken.

Conversion policies include tax incentives such as accelerated depreciation, increased investment tax credit, or guaranteed loans for processors. All these policies are being considered. End use policies include an exemption of excise taxes on gasohol, a production tax credit for alcohol, or increased gasoline taxes on non-alcohol blended fuels.

Energy transition policies

The United States is in the midst of a transition from oil and natural gas as our primary energy sources to dependence on a range of other sources including biomass. Even though the price of oil has increased ten fold in the past seven years, many of the alternative energy sources are still too expensive to be commercially viable without government subsidy. Much of the legislation before the US Congress and the Executive today provides subsidies for development of alternative energy sources.

One possible problem with massive subsidies for emerging energy technologies or sources is that we become prematurely locked into a technological or raw material path which later turns out to be less desirable. This has been called the "foreclosed futures" problem.

With respect to alcohol from grains, the problem could arise if 1) it does turn out in the future that world grain supplies are inadequate or 2) large numbers of grain alcohol plants are built and the cellulose conversion technology subsequently is proven technically and economically. One way to help solve this problem is to require that all grain alcohol plants built with government subsidies be compatible with future conversion to cellulosic feedstocks. Compatibility would require 1) that the plants be planned and constructed so that later conversion to cellulosic feedstocks would be facilitated, and 2) that the plants be located in an area with adequate cellulosic feedstocks available to run the plant completely on these raw materials. These requirements generally would not be difficult to meet, and they would ease the transition to alternate feedstocks in the future. By following this course, constructing grain alcohol plants today would not foreclose shifting to cellulosic feedstocks in the future. To the contrary, it would establish a physical plant and technical base for future implementation of the cellulosic processes.

SUMMARY AND CONCLUSIONS

Biomass energy has the potential of being quite significant in the United States energy picture by the year 2000. Up to 16 quads of biomass energy could be available by 1995-2000. That amount is slightly greater than the 1979 total US consumption of coal and about 43 percent of 1979 oil consumption. The lower end of the biomass range for 1995-2000 is 7 quads which is almost half of the 1979 US coal consumption. Of this total biomass energy potential, the potential from the agricultural sources—grains, forage crops, and crop residues—ranges from 1.9 to 5.7 quads or one-fourth to one-third of the total biomass potential. The potential from wood, the largest single biomass source, is 5 to 10 quads. The US energy potential from grains is 0.3 to 0.7 quads or about 4 percent of the total biomass potential.

Despite the fact that grains are only a small portion of the total biomass potential, a significant amount (3.6–8.4 billion gallons or 13.6–31.8 billion liters) of fuel alcohol could be produced from grains in the United States. This alcohol production would be sufficient to substitute gasohol (10 percent alcohol) for about one half of the 1979 US gasoline consumption. However, this alcohol production could cause significant increases in corn prices. In one simulation result at the 4 billion gallon level, corn prices by 1985 were $3.85/bushel as compared with $2.77 in the base (no alcohol) case, an increase of almost 40 percent.[20] A price increase of this magnitude would cause a reduction in corn exports and eventually a change in the dietary composition of consumers of animal products throughout the world. Initially, consumers probably would shift away from beef to more efficient energy converters such as poultry. These same simulation results indicated that alcohol production up to 2 billion gallons would not cause significant price increases. The economics of ethanol production from grain have changed dramatically in the last two years. In early 1978 the US refinery gate price of gasoline was about $.45 per gallon, and the cost of producing anhydrous ethanol was about $1.05 per gallon. In early 1980 the refinery gate gasoline price has more than doubled to about $.95 per gallon while the ethanol price has increased about 15 percent to $1.21. When the octane enhancement effect of ethanol is taken into consideration, grain alcohol is economic or nearly economic today in the United States.

With the improved economics of alcohol production, the increased cost of oil imports, and the strong political support of farm groups, alcohol production policies are receiving a tremendous amount of interest and attention in the US Congress. Alcohol production will have a subsidy of at least forty cents per gallon lasting through 1992. In addition, tax credits, low interest or guaranteed loans, state tax incentives, purchase guarantees, and the like will increase the effective subsidy considerably higher—in some cases to as high as $1.20 per gallon.

At such large subsidy levels, alcohol production in the United States is likely to increase very rapidly in the 1980s. It could easily reach 2 billion gallons per year or higher by 1990. The work done to date indicates that US agriculture could absorb an alcohol demand for grain at the 2 billion gallon level plus the continually increasing domestic and export demands without any major changes in prices. Beyond this alcohol production level, however, major changes may occur, so it appears that US government policies should be designed so that the subsidies can be eliminated or phased out if grain prices rise substantially. Unfortunately, the subsidies now being considered are not linked in any way to world oil prices or grain prices. A major challenge to policy makers in the United States is to provide subsidies and policies that are sufficiently flexible so that they can be modified as economic conditions change. Policy flexibility is essential as we enter this new era of energy production from agriculture if we are to wisely manage the food/feed/fuel trade-offs.

NOTES

1. A quad is one quadrillion BTUs, 1.055 million terajoules, or approximately one-half million barrels per day of oil.
2. Wallace E. Tyner and J. Carroll Bottum, "Food vs. Fuel. The Gasohol Dilemma," paper prepared for the National Alcohol Fuels Commission, Washington, D.C., January 1980, p. 2.
3. Sugar crops are not expected to be used in large amounts for energy in the United States.
4. The forage crop numbers presented here are in addition to the forage used for livestock production.
5. Wallace E. Tyner and numerous other faculty in the School of Agriculture, Purdue University, *The Potential of Producing Energy from Agriculture*, (Washington, D.C.: Office of Technology Assessment, May 1980), pp. 6–109.
6. *Ibid.*,pp. 170–85.
7. W. E. Larson, R. F. Holt, and C. W. Carlson, "Residues for Soil Conservation," ed. W. R. Oschwald (Madison, Wisconsin: Special Publication 31, American Society of Agronomy, 1978), pp. 1–18.
8. Sterling Wortman and Ralph W. Cummings, Jr., *To Feed This World: The Challenge and the Strategy* (Baltimore: Johns Hopkins University Press, 1978), pp. 17–24.
9. Rise is excluded from this study because it is unlikely to be ab important input for alcohol production in the United States.
10. Cooperative Extension Service, Institute of Agriculture and Natural Reources, *Ethanol Production and Utilization for Fuel*, Lincoln: University of Nebraska, 1980), p. 42.
11. Distillers grain also contains about the same energy as corn.
12. Donald Hertzmark and Brian Gould, *The Market for Ethanol Feed Joint Products*, (Golden, Colorado: Solar Energy Research Institute, 1979), pp. 3–12.
13. The 65 bushel per acre assumption was a consensus number reached by several agricultural economists at Purdue.
14. Ronald Meekhof, Wallace E. Tyner, and Forrest Holland, "U.S. Agricultural

Policy and Gasohol: A Policy Simulation," *American Journal of Agricultural Economics*, 62 No. 3 (August 1980), pp. 408–15.

15. Ronald Meekhof, Mohinder Gill, and Wallace E. Tyner, *Gasohol: Prospects and Implications* (Washington, D.C.: US Department of Agriculture, June 1980), pp. 12–16.

16. Office of Technology Assessment, *Gasohol: A Technical Memorandum* (Washington, D.C.: Office of Technology Assessment, September 1979), p. 35.

17. Office of Technology Assessment, *Energy from Biological Processes*, (Washington, D.C.: Office of Technology Assessment, July 1980), p. 6.

18. D. Gale Johnson, *World Food Problems and Prospects* (Washington, D.C.: American Enterprise Institute, 1975).

19. This statement is not intended as a statement of official US policy; rather, it is this author's interpretation of US policy.

20. Meekhof, Tyner and Holland, "*U.S. Agricultural Policy and Gasohol: A Policy Simulation.*"

BIBLIOGRAPHY

Cooperative Extension Service, Institute of Agriculture and Natural Resources. *Ethanol Production and Utilization for Fuel*. Lincoln: University of Nebraska, 1980.

Hertzmark, Donald, and Gould, Brain. *The Market for Ethanol Feed Joint Products*. Golden, Colorado: Solar Energy Research Institute, October 1979.

Johnson, D. Gale. *World Food Problems and Prospects*. Washington, D.C.: American Enterprise Institute for Public Policy Research, 1975.

Larson, W. E.; Holt, R. F.; and Carlson, C. W. "Residues for Soil Conservation." In *Crop Residue Management Systems*, pp. 1–18. Edited by W. R. Oschwald. Madison, Wisconsin: American Society of Agronomy, Special Publication 31, 1978.

Meekhof, Ronald; Gill, Mohinder; and Tyner, Wallace. *Gasohol: Prospects and Implications*. Washington, D.C.: United States Department of Agriculture; Economics, Statistics, and Cooperatives Service, Agricultural Economic Report No. 458, June 1980.

Meekhof, Ronald L.; Tyner, Wallace E.; and Holland, Forrest D. "U.S. Agricultural Policy and Gasohol: A Policy Simulation," *American Journal of Agricultural Economics*, 62 No. 3 (August 1980), pp. 408–15.

Office of Technology Assessment. *Energy from Biological Processes*. Washington, D.C.: Office of Technology Assessment, US Congress, July 1980.

Office of Technology Assessment. *Gasohol: A Technical Memorandum*. Washington, D.C.: Office of Technology Assessment, US Congress, September 1979.

Organization for Economic Co-operation and Development. *Interfutures, Facing the Future*. Paris: Organization for Economic Co-operation and Development, 1979.

Raphael Katzen Associates. *Grain Motor Fuel Alcohol—Technical and Economic Assessment Study*. Report, Washington, D.C.: US Department of Energy, June, 1979.

Rask, Norman. "Food or Fuel? Implications of Using Agricultural Resources for Alcohol Production." Manuscript, Columbus, Ohio: Department of Agricultural Economics, Ohio State University, 1979 (photocopied).

Tillman, David A., Sarkanen, Kyosti V., and Anderson, Larry L. *Fuels and Energy from Renewable Resources.* New York: Academic Press, 1977.

Tyner, Wallace E., and Bottum, J. Carroll. *Agricultural Energy Production: Economic and Policy Issues*, Station Bulletin No. 240. W. Lafayette, Indiana: Department of Agricultural Economics, Purdue University, 1979.

Tyner, Wallace E., and Bottum J. Carroll. "Food vs. Fuel: The Gasohol Dilemma?" Report, Washington, D.C.: National Alcohol Fuels Commission, January 1980.

Tyner, Wallace E., and numerous othe faculty of the School of Agriculture, Purdue University. *The Potential of Producing Energy from Agriculture.* Report, Washington, D.C.: Office of Technology Assessment, US Congress, May 1979.

United States Department of Agriculture, *Agricultural Statistics 1972.* Washington, D.C.: US Government Printing Office, 1972.

United States Department of Agriculture, *Agricultural Statistics 1978.* Washington, D.C.: US Government Printing Office, 1978.

United States Department of Agriculture; Economics, Statistics, and Cooperatives Service. *Feed Situation.* Washington, D.C.: US Government Printing Office, August 1979.

United States Department of Agriculture; Economics, Statistics, and Cooperatives Service. *Wheat Situation.* Washington, D.C.: US Government Printing Office, August 1979.

United States Department of Agriculture; Economics, Statistics, and Cooperatives Service. *Foreign Agricultural Trade of the United States*, Washington, D.C.: US Government Printing Office, 1979.

United States Department of Energy, *The Report of the Alcohol Fuels Policy Review.* Washington, D.C.: US Government Printing Office, June 1979.

Wortman, Sterling, and Cummings, Ralph W. *To Feed This World: The Challenge and the Strategy.* Baltimore: Johns Hopkins University Press, 1978.

PART VIII

Opportunities in the Transition Phase

Chapter 18

International Energy Issues: the Next Ten Years

Professor Peter R. Odell[*]

INTRODUCTION

We have recently finished a simulation study on the inter-relationships of the three main components which determine the global future for oil, viz. the size of the resource base, the rate of annual additions to reserves and the growth (decline) in the annual use of oil.[1] From this study we conclude that there is a 90 percent probability that the industry—world wide—could continue to grow until the year 2011; a 50 percent chance that growth in oil production could continue until 2033 and a 10 percent probability that the industry need not reach its peak until 2072. Over the remainder of the industry's growth period the world-wide use of oil could increase to at least double and up to 3.3 times its present size (of 23.6×10^9 barrels annual production) and even after its peak the industry would have several decades to go before it was back to its present level of development—to 2039, 2077 and to well after the year 2080 (the end date of the simulation) in terms of the 90 percent, the 50 percent and the 10 percent probability levels, respectively.

---

*Erasmus University.

THE DEVELOPMENT OF A SUPPLY
CRISIS SYNDROME

These conclusions are markedly at variance with the accepted wisdom of the moment as far as the future of oil is concerned. According to such wisdom the world faces an inevitable, near-future scarcity of oil.[2] It must also be stressed, however, that the conclusions are equally at variance with the accepted wisdom on the future of oil up to less than a decade ago. There was then a general concensus as to the ability of the oil industry to go on expanding at the exponential rate it had achieved over the previous 25 years. This was presented as an outlook for oil which would last far longer into the future than one really needed to think about.[3]

In our study we have attempted to demonstrate the inherent irrationality of such a supremely optimistic view on the future of oil even in the context of the largest ultimate oil resource base (conventional and unconventional) which we have been persuaded it would be wise to consider, viz. $11,000 \times 10^9$ barrels. It was, nevertheless, on the basis of that general concensus view of a guaranteed future for oil in the long term that the world embraced an increasing dependence on oil. In that period of rapidly expanding oil availability and use, the development possibilities of other energy sources were undermined and reasonable attitudes concerning the manner of oil use were overwhelmed by the belief in the desirability and inevitability of a 7½ percent per annum exponential growth. In other words, policy decisions for the short term were, from the mid-1950s to the early 1970s (a period of almost 20 years), based on an inappropriate interpretation of the longer term prospects for the availability of oil. It is, in essence, from that earlier complete failure to appreciate the complex and dynamic interrelationships between oil resources, reserves and use that the world's current energy problems—the so-called energy crisis—really stem. In the first place we were completely disarmed in respect of the immense dangers inherent to the system from a ten-year doubling rate in the demand for oil. These arose not only because of the impact of such a development on the rate at which new resources of oil had to be found, but also because of the potential power that it gave to the oil producing countries.[4] Secondly, it also produced a powerful disincentive for countries and consumers to use even known energy resources which were adjudged to be just slightly higher cost than the low cost oil of the prolific oil-bearing regions which were discovered and exploited in the 1950s and the 1960s. There was certainly no incentive to look for new energy resources—including undiscovered oil resources in non-producing countries—in most other parts of the world.

Thus, the world's increasingly oil-intensive economies and societies became dependent on oil reserves which were unduly concentrated in a relatively small number of countries. This was most marked, of course, in respect of the Middle East where, for reasons which originate essentially from the political relationships of the world's Great Powers in the nineteenth century and the early part of the twentieth century, the oil industry had been strongly motivated to concentrate its attention. This region of the world, in other words, not only appeared to be unique in its geological characteristics as a habitat for oil, but it has also seemed to be well-nigh unique in politico-economic terms. It achieved the status of the happy hunting ground for the Anglo-American oil companies which, until very recently, were able to secure conditions there for the exploitation of oil which were of a kind generally unattainable throughout most of the rest of the world.

It is thus 'accidents' of recent political and economic history which have come to determine the geographical pattern of the non-communist world's reserves of oil. This has had a profound effect on the present outlook on oil in that such a concentrated pattern of proven reserves has produced the belief that this pattern necessarily represents the ultimate geography of oil resources. Given the present situation in the Middle East, this implies great uncertainty concerning the near-future availability of the oil that has been discovered and exploited. It is from such concern that there evolves the oil scarcity syndrome with which we have become familiar in the last few years.

The international oil companies failed earlier to secure their acceptability in many parts of the world as a result of factors of nationalism and in light of the view that the oil companies were the main agents of economic imperialism. The companies, more-over, have also been upstaged since 1973 by the oil producing and exporting countries (O.P.E.C.), as a result of which they have lost their control over most of the world's proven oil reserves as well as over the potential oil resources in countries whose oil industries they had created. This change in ownership and control over a commodity essential to the west's modern economic and social system has produced the crisis of confidence over the future of oil.

Unhappily, it is in this atmosphere of crisis, arising out of the failure of the international oil companies, in their role as economic and political institutions, to maintain their position viz à viz the major oil-exporting countries, that policy decisions are now being generated which, in essence, have the effect of exacerbating the crisis. Indeed, they serve to turn it from a short-term crisis which simply necessitated a limited period of readjustment, by both the suppliers and the users of oil, to one in which the future of oil has been heavily discounted in spite of the fact that it is the energy source most appropriate for continuing to make both economic and social developments possible.

In other words, what are essentially short-term difficulties over oil, arising out of geo-political changes, have become hopelessly interwoven with essentially separate longer term considerations. The shortage of oil which, in fact, emerges out of the shift in the balance of oil power (from the international companies to oil-exporting countries) is now being incorrectly interpreted as a necessary long-term condition of the world's economic outlook. As such it is also producing policy decisions and actions which themselves serve to ensure that the very scarcity which is feared is, indeed, made a permanent condition. However, as we have demonstrated by our simulation study on the future of oil, scarcity is un-necessary for a period of *at least* thirty years whilst, at the other extreme, it may not be a necessary development, from the point of view of the dynamics of resources, reserves and demand inter-relationships, for another 100 years.

THE COST OF OIL

Acceptance of the hypothesis that a near-future physical scarcity of oil is not a necessary condition for the evolution of the world's economic, social and political order, opens up a range of energy options which are otherwise excluded. From the point of view of the cost of energy it casts doubts on the belief that there is only one way the price can go, namely 'up', even from the dizzy heights that have already been secured by the countries currently controlling the supply, and hence the price of oil. Hard though adjustment to $35 per barrel oil may be amongst consuming nations and consumers there is, nevertheless, currently a feeling of inevitability about it. This emerges from the widely held view that the rapid depletion of the world's limited availabilities of oil justifies such a price. The same attitude, moreover, means acceptance of the need to face the relatively near-future and very costly prospects of even more expensive alternative energies—notably nuclear power, the so-called attractiveness of which is basically a function of the belief in the scarcity of oil.

There are even serious doubts expressed over the validity of what appears to be a highly reasonable proposition on the evolution of alternative energy costs, viz. that it is quite reasonable, in the light of historical experience, to hypothesize a fall over time in the real costs of alternatives to conventional oil. Where such alternatives are already in development, progress can be made along the learning curve towards lower technological costs and economies of scale can be increasingly realized as the opportunity to move to bigger units is generated. Fisher[5] has shown that these factors have always worked in respect of the historical processes of energy resources development. Indeed, the essential requirement for the successful development of alternatives to conventional oil is time. Time is required for the learning process, as well as for the achievement of technological break-throughs, on the basis of which reduced costs can be achieved. The real, as opposed to the feared, prospects for oil not only give this necessary breathing space, but they also provide the opportunity for a smooth transition to alternative energies at unit prices which, in real terms, could fall to well below today's $35 per barrel oil.

DISINCENTIVES TO OIL SUPPLY

In the meantime, except in the context of a world supply of the commodity which is controlled by a limited number of countries with production to spare for export, $35 per barrel is well above the long-term supply price for conventional oil production. This is true almost irrespective of its habitat (including deep-water offshore), of its reservoir characteristics (in terms of size, and depth), and of difficulties in recovering it as a result of specific physical and chemical variables in its occurrence. Revenues of $35 per barrel available in total to the potential producing company would give a return on investment more than adequate to justify a 'go' decision in almost every case.

However, in the context of the general belief in oil as a scarce commodity and given the ability of governments to lay effective claim to all or most of the economic rent emerging from the production of relatively low-cost oil and its sale at high prices, the producing companies cannot lay their hands on more than a small proportion of the revenues generated by $35 per barrel oil. Governments' claims to most of the economic rent from oil production emerge from the idea that the depletion of a country's 'scarce' resources can be justified only in terms of the revenues they earn for governments. On the basis of such government revenues the 'oil can be sowed' (to use the Venezuelan term) to diversify and strengthen the economy. The consequence of such attitudes and actions is, however, to eliminate a greater part of the motivation for companies to seek more oil, and/or to produce more of that which they have found. This is another element in the man-made process of creating a scarcity of oil.

In such circumstances it is hardly surprising that the oil companies prefer to diversify into other sorts of energy which, though inherently more expensive to produce, usually give the prospects of a better return on investment as governments, to date at least, have not had the same propensity to intervene in their exploitation and development. Or, even worse, from the point of view of the world's energy outlook, the companies prefer to diversify their investments away from energy altogether and into activities in which governments seem even less likely to get involved. This, of course, works to the detriment of the potential supply of the energy that may be required to complement the now slowly increasing need for oil.

Thus, unhappily, not even $35 for a barrel of oil in 1980 necessarily enhances the supply potential. Thus, as a consequence of the behaviour of most governments—both in terms of their belief in oil as an inherently scarce commodity and in the light of their attitude that oil is a commodity the production of which can be used to ensure a flow of government revenues on a scale hitherto undreamed of—the supply of oil thus seems likely to remain much less elastic than might reasonably be concluded, were this to be viewed simply as a proposition in economics.

Moreover, this is not the only, or even the most important, way in which government intervention inhibits the development of a supply of oil in the quantity which could otherwise be expected with oil at its present price. There is also the question of governments' price expectations and the low discount rates they tend to use in order to compare present with future alternatives. If oil is scarce (again the first and basic assumption of the analysis in which energy planners indulge), so that the price will rise over time, then how much easier it is to demonstrate that it is preferable to keep the oil in the ground, rather than to produce it. Hence, the search for, and the hope of the achievement of, optimum national oil depletion rates by governments as dissimilar as those of Kuwait, Britain, Canada, Norway and many others in different parts of the world. The end results of this factor are: first, less production in the immediate future (hence exacerbated short-term scarcity and consequential politico-economic problems of the kind which now so beset the western economic system); second, less development of known oil provinces and hence the propensity to create a medium-term scarcity of oil); and, third, less exploration (with its inevitable consequences; viz. the achievement of a long-term scarcity of the commodity, simply because potential supplies have not been intensively enough sought).[6]

And there is still another set of forces at work, in the context of national energy planning activities, which also serves to help make the belief in scarcity more of a reality. This is the set of forces concerned with the feared adverse effects of oil exploration and exploitation on the environments of particular countries. It thus includes concern for pollution—both of water and of atmosphere; concern for the social effects of oil developments on specific communities and on particular localities; concern for the impact of oil developments on other economic activities which may have to compete with the oil industry for scarce resources such as labour; and, finally, concern for the macro-economic effects of oil export earnings and of oil revenues on the structure of a country's economy. These, however, are all basically management and control problems and none of them are incapable of solution in sophisticated western industrialized societies, even with rapid and extensive oil exploration and exploitation activities. They are, nevertheless, problems which can be, and, indeed, sometimes are, especially in Western Europe, used as reasons for restricting the rate at which oil developments are permitted. Unhappily, in the enthusiasm, often by particular vested interests, for the short-term justification for such restrictions, the longer term issue of the cumulative effect that they have on the outlook for oil is usually ignored or, if treated at all, it is relegated to a minor role in the decisions. The consequence is a contribution to the creation of oil scarcity in the medium to the longer term.

THE IMPACT OF THE STRUCTURE OF THE INDUSTRY

It is, however, not only the governments of oil producing countries (or of potential producing countries) which contribute to the likely achievement of a scarcity of oil in the long, as well as in the short, term and to the creation of a higher-than-necessary price for oil as a problem of the moment. The future development of the future supply and price of oil is also a function of the structure, the organization and the financing of the industry. As we have mentioned earlier in the chapter, the international oil companies, which have hitherto been responsible for most technological progress and most investment in oil, have become decreasingly acceptable as appropriate instruments for the exploitation of oil in many parts of the world. This, in the short to medium term, inevitably limits the degree to which new oil can be found and produced and the extent to which already discovered and producing fields can be further developed with enhanced recovery facilities. This is a function

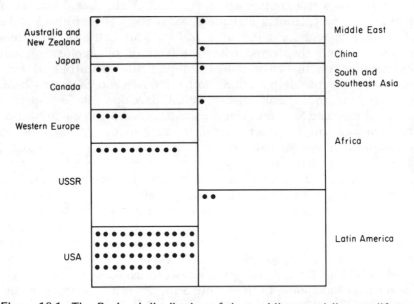

Figure 18.1. The Regional distribution of the world's potentially petroliferous areas: shown in proportion to the world total. Within each region the number of exploration and development wells which have been drilled is shown—each full circle represents 50,000 wells. Relative to the United States all other parts of the world—but especially the regions of the Third World—are little drilled for their oil.

of the fact that the expertise remains concentrated in the hands of these companies and that there has been inadequate development to date of any really large-scale and effective alternative to the companies for the exploitation of most of the world's oil resources. In addition, even in the areas of the world in which the companies continue to work they do not necessarily find it commercially worthwhile to produce oil as intensively or as extensively as the resource base allows. In the newly developed North Sea oil province, for example, many fields with less than 200 million barrels of estimated recoverable reserves, together with parts of many larger fields, whose oil cannot be recovered through a production system which gives an adequate return on investment to the company concerned, have been left unexploited because the companies responsible for their discovery/ development have determined that the investment necessary to produce their oil would not generate a high enough rate of return to justify its being made.[7]

Thus, in areas and conditions such as this—and particularly in large

Table 18.1. Estimates of the Ultimate Oil Resources of the Third World.
(in bbls × 10⁹)

	Oil Industry Views (a)	Grossling (U.S.G.S.) (b)	Min. of Geology USSR (c)
Latin America	150-230	490-1225	620
Africa	120-170	470-1200	730
South/S.E. Asia	55- 80	130- 325	660
Totals	325-480	1090-2750	2010

Sources: (a) Based on figures in R. Nehring, *Giand Oil fields and World Oil Resources,* Rand, Sante Monica, June 1978. Table on p. 88 adjusted to give comparable geographical coverage with (b) and (c). Nehring also states (p. 88) that his figures are "roughly similar" for these regions to those published elsewhere by oil industry observers. (b) B.F. Grossling (USGS), "In Seach of a Probabilistic Model of Petroleum Resources Assessment" in *Energy Resources,* M. Grenon, Ed., I.I.A.S.A., 1976. (c) Visotsky et al, Ministry of Geology, Moscow, "Petroleum Potential of the Sedimentary Basins in the Developing Countries", *ibid.*

parts of the Third World where, as Figure 18.1 shows, much of the world's remaining oil potential lies—the future development of the world's oil resources depends to a considerable extent upon a changed structure for the organization of the oil industry. Without such a change, there appears to be little chance of ensuring a large enough flow of expertise and investment into oil exploration and development in order that the world's oil resource base—whether of 3,000 or 11,000 × 10⁹ barrels— can be adequately exploited over the next 30 to 100 years.

There is thus a further basic paradox in the world of future oil potential. The oil companies, imbued with a sense of political realism that arises out of their long experience in working in most countries of the world, now discount their opportunities to exploit much of the remaining potential. In their public presentations of this 'fact of life', seen from their point of view, they help to create and sustain the idea of oil as a scarce commodity. This may be seen, for example, in their views on the limited potential resources of Latin America, Africa and South East Asia. These views may, however, be contrasted with those of other organizations (*see Table 18.1*). But because the companies, in spite of the hostility which is often shown towards them, are viewed generally as the best source of information (and even as objective sources of information) on the prospects for oil, then their commercially inspired interpretations are accepted and, in so being accepted, they enhance the inherent scarcity belief.

In other words, in spite of their undoubted expertise and increasing technological capabilities both in finding and producing oil, the companies are responsible for much of the pessimism and the lack of confidence that now pervades the international scene in respect of the future of oil. Yet, as we have argued and as the companies themselves so recently stressed, such pessimism and lack of confidence is by no means a necessary part of the outlook for oil; and as oil remains the main source of energy for a developing world this is an important issue. The massive swing in the companies' attitudes to the future of oil—from the supreme optimism and confidence of the period up to the oil crisis to their present pessimism and lack of confidence—presents a sobering contrast. It can be shown that their earlier extremely optimistic views were unwarranted. There is also much evidence to suggest that their new, extremely pessimistic views are just as likely to be unjustified.

THE FUTURE OF OIL

Thus, the evolution of a future of oil which could relate closely to the statistical probabilities emerging from the inter-relationships of resource base size, adequate additions to discovered reserves year after year, and to a modestly increasing, and increasingly efficient, use of oil in the development process[8] depends upon three inter-related factors. First, the elimination of the present crisis mentality in respect of the longer term availability of oil. Second, the willingness of governments, particularly those of the O.P.E.C. and of the O.E.C.D. groups of countries, to sustain and encourage the exploration and exploitation of their countries' oil resources. And third, the establishment of an adequate organizational infrastructure to make the required exploration and exploitation efforts possible in all parts of the petroliferous and potentially petroliferous world.

Important though the first two of these factors are for the future of oil, especially in the short to medium term, it is the third factor which appears to constitute the critical variable for the longer term outlook. Indeed, any success achieved in the establishment of an infrastructure appropriate for the future exploitation of all the world's oil resources would automatically generate conditions in which changes in attitudes to the availability of oil and in government policies towards its development could be expected almost as a matter of course.

The understatement of the world's oil potential is, as we have stressed already, currently justifiable from the point of view of the international oil companies. Such commercial justification for the contemporary understatement arises in exactly the same way that over-statement on the companies' part was justified when their commercial viability depended on their success in selling rapidly increasing quantities of oil in the context of energy markets in which there was plenty of scope for oil to substitute other sorts of energy (for example, coal in Western Europe and the United States and vegetable fuels in Latin America).

In today's changed international economic and energy market conditions, however, the effective resource base for the companies is a more important consideration. And for them the effective base is the one they expect to be able to exploit. If their relations with many countries of the world are either so bad or non-existent that the companies cannot conceive of circumstances in which they could commit large investments to oil developments to them, then the companies have no option but to discount the resources concerned— and so visualize the world's energy future as one in which the output of oil will soon reach its peak and thus need to be supplemented by the use of coal and nuclear power etc. This is the kind of view which Shell and B.P. have recently tried to put across in their public relations' type advertising.

However, the fact that the western world currently depends on institutions which, by virtue of their very nature, cannot find and develop much of the world's oil, does not mean that the resource base simply cannot be developed. The international oil companies constitute a set of institutions which is capable of being changed, and/or substituted and/or complemented. New or revised institutional arrangements for tackling the opportunities provided by the under-explored and little developed parts of the world are capable of being created—within a decade or so. This development would be in time to take up the exploration for, and the exploitation of, the world's oil prior to the reserves/production ratio falling to such a low level that confidence in the future of oil is completely undermined. Our simulation of the 50 percent probable future of oil shows that it will be 1995 before the R/P ratio falls to the level at which the world oil industry needs to find more reserves. There is thus a breathing space available for action to be taken which extends even beyond the minimum likely period necessary for the creation of a revised institutional framework capable of ensuring the long-term development of the industry.

A NEW ROLE FOR THE COMPANIES

Given that successful oil development depends on the continuing availability of managerial and technological expertise and on large-scale financial resources; and given that 90 percent—or possibly more—of such resources exist within, or are only available to, the international oil companies and a limited number of other entities, especially in the United States, then an expanded involvement of these companies in the future of oil would appear to be a prerequisite for success. For these companies to give less attention to, or for them to be forced to give up, oil exploration and production could well be the means whereby an otherwise achievable long-term future of oil is undermined.

Thus, the future of oil may well depend on finding a way in which the international oil companies can become associated with exploration and production in as many countries of the world as possible. This, of course, means that their involvement has to be politically acceptable to the countries concerned, whilst from the companies' point of view it has to be commercially attractive.

Somewhat paradoxically, in the light of the recent nationalisation of their oil industries, it is in the traditional oil producing and exporting countries that the need to re-involve the oil companies has already been recognized and, indeed, already acted upon. This has been done as a means whereby the countries can secure access to the companies' expertise, even if not to their financial resources as these hardly constitute a problem for the member countries of O.P.E.C., with oil at $35 per barrel. In many O.P.E.C. countries agreements have been signed with the oil companies whereby the latter, in return for a fee (usually related to production levels) offer managerial and technological knowledge and services to the state oil company, so that it will be in a better position, first, to ensure the flow of oil from fields that are already in production and, second, to ensure the continued discovery of new fields and of new oil producing regions for the longer term future of the industry of the country.

Such arrangements avoid the earlier political unacceptability of the companies and provide a way whereby their accumulated expertise can be tapped. The fact that major oil producing and exporting countries can negotiate this sort of agreement is an indication of their new-found politico-economic strength and of the confidence they have in their ability to control the situation in their own interests. The companies, for their part, achieve contracts which are virtually riskless in that payments are related to existing levels of production and are due, moreover, from countries for which government revenues pose no problems.

It is, however, difficult to visualize that this sort of country/oil company relationship could develop quickly in respect of the opportunities for oil exploration and development in the world's less developed countries. There are several reasons for this. First, the necessary parity of esteem does not exist between the two parties to enable effective managerial/technological relationships to be established. Second, most less developed countries lack the necessary knowledge of the oil industry to handle the complex negotiations with confidence. Third, the oil companies have insufficient confidence in the opportunities offered by such countries, most of which lack the necessary infrastructure to make oil activities easy to develop. Fourth, and most important, the countries concerned usually lack the very necessary and the very considerable financial resources to invest in the high risk business of finding and developing oil. The oil companies are, of course, aware of this and thus rate as rather low their chances of being paid for efforts which turned out to be unsuccessful in finding oil.

INVESTMENT IN THE FUTURE OF OIL

A recent World Bank report has very conservatively indicated an immediate potential availability of at least 60×10^9 barrels of oil in those parts of the Third World in which the international companies, for the reasons set out above, find it difficult or impossible to work.[9] The report indicated the opportunity to develop a production of over 5 million barrels per day within a 10 year period, given a level of investment in exploration, production and associated transport facilities which, in 1976 dollar terms, would require upwards of $6,000 million per year in investment. Needless to say, in relation to this size of annual financial requirement, the World Bank also indicated that it could not help, except in a very minor way. It would not be able to lend money for exploration because it is too risky, and even in terms of investment for field development and transport facilities the World Bank would be limited to undertaking two projects per year with each involving only $300 million. This is, in essence, an indication of a near zero ability by the western world's main international financing agency to help develop the Third World's oil resources.

There is thus a formidable inability on the part of existing private and public international institutions to do anything significant to ensure the development of most of the world's remaining oil resources, in spite of the fact that the commitment of funds and of other resources to such development seems more likely to produce more energy per dollar invested than the investments made in the search for alternatives. This is true whether the search involves renewed interest in old oil regions, in order to try to recover a little more of the oil in place, or investment in other sources of energy. Such alternative investment strategies are, however, being followed. But their imposition is a result of the inadequate contemporary institutional facilities for finding and developing lower cost oil reserves and it thus represents a grossly sub-optimal allocation of the world's limited financial and technical resources in the search for energy.

Yet this development is not in the long-term interests of any of the parties involved. Both the industrialized (the main oil consuming) countries and the present oil producing (and exporting) countries, as well as the poor Third World countries themselves, have an interest in the creation of a system in which the world's oil resources can be sought and developed at a rate which is high enough to sustain a growth in use needed to keep the world's economy moving ahead. The responsibility for ensuring this is shared by the two first named groups of countries whose joint ability to provide a level of funds capable of sustaining the required effort is a matter of political will, rather than of economic ability. Given their co-operation, perhaps under the aegis of a joint O.E.C.D./O.P.E.C. agency, the international oil companies and other appropriate institutions could be hired to manage and/or to undertake oil exploration and development work in Third World and other countries.[10] As a result, the companies would, on the one hand, be "de-politicized". They would be hired simply for a specific exploration and development effort in the context of which they could not exercise any political or economic power and influence in the countries concerned. On the other hand, the companies themselves would be confident of being paid for their services and so be able to judge the requests made to them in a normal commercial way. The end result would be a continuing and a geographically much more dispersed pattern of oil exploration and development. This, on being implemented over the long term, could ensure a future of oil more closely approximating to the opportunities offered by the world's ultimate resource base, and so give a 90 percent probability that oil will remain a growth industry into the second decade of the twenty-first century and a 50 percent probability that oil will be able to meet much of the world's modestly growing needs for energy until almost the middle of the next century.

And, in the meantime, those countries with oil reserves and/or the prospects for rapidly enhancing those reserves through the further exploration of known oil provinces might become more willing to increase their levels of production and thus serve to diminish the spectre of widespread economic and social difficulties arising from the shortfall in oil supplies over the next five to ten years.

NOTES

1. P. R. Odell and K. E. Rosing, *The Future of Oil: a Simulation Study of the Inter-relationships of Resources, Reserves and Use, 1980-2080*, Kogan Page, London , 1980 (forthcoming).
2. See, for example, B.P.'s *Oil Crisis . . . Again?*, (London, 1979) and D. dr Bruyne (President, Royal Dutch Petroleum Company), "Energy Imperatives for the Coming Decades", in: *Energy, What Now?*, (K. E. Davis and P. de Wit, Eds.), B. V. Uitgeverij Bonaventura, Amsterdam, 1979.
3. See, for example, "Petroleum and the Energy Crunch", an editorial in the *Oil and Gas Journal*, November 15, 1971; J' M. Moody (Mobil Oil Corporation), "Petroleum Demands of Future Decades", *American Association of Petroleum Geologists Bulletin*, Vol. 50, No. 12, December 1970; A. Hols (Shell), "Oil in the World Energy Context", *E. I. U. International Oil Symposium*, London, October 1972; J. D. Parent and H. R. Linden, *A Study of Potential World Crude Oil Supplies*, Chicago, 1973; Anon, *Petroleum Economist*, Vol. XXXIV, No. 10, October 1972.
4. The authors made this point, and attempted to quantify the potential costs involved, in an article published as early as February 1971. See P. R. Odell, "Against an Oil Cartel", *New Society*, No. 437. 11 February 1971.
5. John C. Fisher, Energy Crises in Perspective, J. Wiley and Sons Inc., New York, 1974.
6. This multi-faceted oil play effectively describes the state of the game in Western Europe where known reserves are deliberately not produced; where oil and gas provinces are deliberately restricted in terms of their development, and where exploration is deliberately discouraged. And yet governments of Western European countries express their deep concern over the prospects for future supplies of oil. For a further discussion of this paradoxical situation see, P. R. Odell, "Towards a More Rational View of the Energy Policy Options open to the Western Oil Consuming Nations, in: *Energy—What Now?*, K. E. Davis and P. de Wit (Eds.), Bonaventura, Amsterdam, 1970.
7. See P. R. Odell and K. E. Rosing, *The Optimal Development of the North Sea Oilfields*, Kogan Page, London, 1976, for an extended discussion of these issues.
8. We shall not discuss the question of the efficient use of oil any further in

this paper. Other studies show the immense scope that exists for so enhancing the efficiency of oil use that the total used need not increase above today's levels even in a growth economy. See, for example, G. Leach et al., *A Low Energy Strategy for the United Kingdom*, I.I.E.D., London, 1978.

9. World Bank Report no. 1588, *Minerals and Energy in the Developing Countries*, Washington, May 1977.

10. The establishment of such an Agency, in the context of the new international oil situation and the potential for its development, has been fully discussed and evaluated in P. R. Odell and L. Vallenilla, *The Pressures of Oil*, Harper and Row, London, 1978.

PART IX

Risks and the Need for Contingency Planning

Chapter 19

Managing Oil Contingencies: the United States Experience and Choices for the Future

Michael L. Telson**

In this paper I will discuss the US experience with managing oil contingencies, and future alternative approaches for dealing with them. I will posit that:

1. world oil market insecurity is a basic fact of life for the United States over the indefinite period ahead; that
2. learning to deal with possible oil contingencies in as efficient a manner as possible is crucial to the economic performance and political stability of the United States; that
3. the United States has two broad strategic choices for the management of future oil contingencies: it can choose to use market mechanisms for allocating scarce supplies, or it can instead increase its use of non-market mechanisms to allocate these supplies; and, finally, that

*Energy Analyst for the Committee on the Budget, US House of Representatives.

**The author is indebted to many colleagues for valuable discussions, suggestions, and critiques; among them, especially Joel Eisenberg and Richard Morgenstern should be mentioned. However, the views expressed in this paper are only the author's, and do not necessarily reflect those of the Committee on the Budget, or any of its members.

4. with the October, 1981, termination of the allocation and price controls authorized by the Emergency Petroleum Allocation Act of 1973, these two alternative approaches will be the subject of increasingly lively discussion in the United States throughout the year ahead.

POSSIBLE OBJECTIVES OF A STRATEGY TO MANAGE OIL CONTINGENCIES

I submit that there are three broad objectives that could be addressed by a strategy to manage oil contingencies:

1. a market efficiency objective,
2. the plan should be viewed as generally equitable to a substantial part of the population, and
3. that the oil industry market structure be given some degree of protection in the event of an oil contingency.

I will suggest in my remarks that market techniques could play an invaluable role in managing oil contingencies, and that, under most circumstances, they would better satisfy the objectives I just listed than the allocation/price control system the United States now has.

THE NEED FOR A STRATEGY. INSECURITY OF THE WORLD OIL MARKET IS A BASIC FACT OF LIFE

I should begin at the beginning. The fundamental problem for the United States and oil consuming country energy policy has been—and over the indefinite future will continue to be—the insecurity of the world oil market. So far, we have lived through two very expensive episodes: the first occurred in 1973, as the consequence of Saudi Arabia's deliberate act to curtail its production of oil; and the second in 1979, when Iranian oil production went down as a largely incidental result of the political turmoil in that country. In 1980 and beyond, the threats will come from both deliberate actions and unanticipated accidents. Roughly 18 million barrels a day (MMBD) out of the 30 MMBD in world trade come through the Strait of Hormuz, and an even higher proportion is produced in very politically unstable areas. It is difficult to foresee how the world could avoid a future interdiction of at least part of that oil.

The response of the United States to this kind of situation, to my mind, should be fairly obvious in a general sense. For the long term, the United States must work to make the world oil market a more stable one—one that would ideally have substantial excess oil production capacity in countries which would like to produce and sell that oil if market conditions would permit. I should be very clear on this point: additional producing capacity in Saudi Arabia will not solve the market insecurity problem, instead, it will make the market that much more dependent on Saudi Arabia's production decisions.

For the long term, then, the United States must strive to reduce its imports of oil, while seeking to stimulate directly additional excess oil producing capacity abroad. It can do the former by making increasingly efficient domestic use of oil, while increasing its production of domestic resources. It can do the latter by seeking to encourage its exports of coal, and of energy conserving technology, while helping lesser developed countries to discover and develop new oil and other energy resources. Now, additional oil producing capacity may not necessarily lead to lower prices for oil, but it will go a long way toward making the world oil market more secure, one that is less prone to violent change, or to concerted manipulation by a handful of countries.

But all of these things will take time. And because it will take a long time to change significantly these market fundamentals—if they can ever be changed—the United States must be prepared now to manage any oil contingencies that might occur in as efficient a manner as possible.

REVIEW OF UNITED STATES
EXPERIENCE WITH MANAGING THE
1979 IRANIAN OIL CONTINGENCY

What has been the United States' experience with managing oil contingencies? We have not done well. In 1979, as in 1973-4, we muddled through the shortage, During April through June of 1979, the United States experienced gasoline lines that severely tested our social and political institutions.

It was difficult to explain to Americans why gasoline lines were occurring, even though 1979 oil imports were actually higher than they had been in 1978. It was difficult to explain that this statistic was misleading because oil company inventories had been liquidated in 1978, and were actually being built up in 1979. It was difficult to explain that the lines were occurring when the gross shortage was relatively small certainly less than 10 percent. And it was especially difficult to explain why they were occurring in the United States when no other major country was experiencing them to that extent, and when gasoline supply was short in some cities but not in others, and not short outside of large urban areas in general.

The Allocation System and its Effect on Supply

How did the allocation system react to the oil shortage? It worked generally as follows. As supplies of gasoline went down, each refiner was obligated to give the jobbers and stations he serviced an equal percentage of their base period volumes. Allocations to stations that had opened since the base period further reduced the fraction available to the remaining stations. This was especially bad for areas where old gasoline stations had closed and where new ones were not taking their place. As the old stations closed, the supplies that had been allocated to them were either returned to the national pool, or were kept by the jobber involved; in any event, there was no obligation to return that gasoline to other suppliers in that area.

The imbalance was further aggravated by the fact that the government's first response to the shortage was to create several categories of use which had unlimited access to whatever supplies they desired. For example, this preference was given to all agricultural users; this further decreased the amounts available for everyone else. Also, to give the states more flexibility to manage the shortages, the states were empowered to allocate 5 percent of the supplies available within each state, through the so-called "state set-aside", and make these available to whichever dealers they pleased, under whatever conditions they set—for example, that the dealers be open on Sundays. Unfortunately, the immediate effect of the state set-aside was to further reduce the quantities available for all other users; and because it took time before the states learned how to release these supplies to the general public in an effective way, the result was to reduce the supplies to the general public even further. As a consequence, it was not unusual to find many gasoline stations being given less than 75 percent of their base period volumes.

Further, the gasoline allocation system was designed to protect the various links in the chain from the refiners down to the final consumer, except for the final consumer himself. It was assumed that if the distribution channels were protected, the final consumer would also be protected as well. This assumption, of course, was false. The clearest example I know of occurred in Connecticut with the "Gasland" chain of service stations.

Over the year since the base period was fixed, the "Gasland" company had decided to close all of its gas stations in Connecticut. Under the rules, it could keep its claim to that gasoline even if it took the gasoline to other stations in another state. Because this company was a relatively large factor in the Connecticut market, its disappearance was strongly felt. This ridiculous situation was later reversed, but not until the damage had been done. The result of all of these fixes was to leave many areas in the country with significantly lower supplies than the aggregate data suggested.

Effect of the Allocation System on Demand

So much for the effect of the allocation system on supply, what about its effect on demand? It appears that the 1979 pattern of demand throughout the country was profoundly affected by the gasoline lines of early spring. The result was that the allocation system was sending gasoline supplies to stations around the country roughly in proportion to the previous year's patterns, when in fact the pattern of demand had been brusquely changed. Because drivers were fearful of not finding gasoline in the weekends of 1979, they tended to stay close to home in the cities where they lived. They did not drive as much outside the large urban areas as they had done in 1978. This meant that supplies in the cities—already inadequate to meet the demand of the previous year—were even less adequate to meet the higher potential city demand in 1979. For example, consider the plight of areas whose gasoline stations may have got 75 percent to 80 percent of last year's supplies while trying to serve a potential demand perhaps higher than 120 percent of last year's demand. Former Secretary of Energy Schlesigner was overheard at a cabinet meeting explaining to the President that the gasoline allocation system was putting gasoline where the drivers weren't.

Therefore, even though the shortage in aggregate was small, the relative shortage in many areas, and specially in many large cities, was very large. It reminds one of the old story about drowning in the stream that was only two feet deep on average. These scattered geographical imbalances between supply and demand set the stage for the gasoline lines.

The Development of Gasoline Lines

When gasoline supplies in a given area were short relative to that area's demand, a degenerative cycle set in. Because there was no way for a gasoline station to increase its profits from increasing its sales volume, and because its prices were fixed at a maximum level under price controls, the stations quickly discovered that the way to maximize profits was to cut costs, which meant staying open fewer hours.

As the stations increasingly did this, a panic developed among motorists who now had to keep higher gasoline inventories in their tanks to protect themselves from the high probability that no stations would be open when they might need additional gasoline. The simple step of increasing gasoline tank inventories had the effect equivalent to a one-time, month-long demand increase of about 7 per cent. Because many motorists could not take the chance of being caught short, they started filling their tanks three or four times as often as they had done previously, further aggravating the gasoline lines. This meant the stations could stay open fewer hours still. The whole situation became rather nasty, with one shooting and many fighting incidents reported throughout this period.

MARKET TECHNIQUES FOR MANAGING OIL SHORTAGES

This is where the use of market techniques to help allocate supplies during a shortage comes in. As long as the allocation system continues, it would seem to me desirable to build into it some market flexibility. I happen to be particularly fond of an idea I like to refer to as "gourmet gasoline", a name coined by Elinor Schwartz.

I believe that much of the desperate nature of the gasoline lines was due to the lack of a legal way to allow motorists who needed a given amount of gasoline, to pay significant premiums to obtain it. Rather than pay with money, motorists were forced to pay with their insecurity and with the time they were willing to spend in lines, sometimes several times a week. Instead, let us suppose that the government allowed a certain amount of gasoline—say, 10 per cent of total supplies—to sell at decontrolled prices. The right to sell this decontrolled gasoline would be a very valuable right, and the government could charge gasoline stations that participated in this program a substantial fee—perhaps a percentage royalty fee which could be determined through public auction. It would seem that the possibility of making increased profits would tempt some of these stations to stay open longer than usual—some would even be open at all times. Of course, the higher the price they charged, the longer they would stay open, this is simple microeconimics.

If this started to happen, it should be possible to considerably reduce the length of the gasoline lines. How? By removing those motorists who would be willing to buy their way out of them because they badly needed a given amount. This would work to their convenience, and help everyone else by making the remaining lines shorter. The people who helped create the lines would now know there was a way—at a high price, to be sure—to buy the amount of gasoline they most needed. The fees collected by the government could be used to finance transfer payments to address the equity issues involved with decontrolling the price of these gasoline supplies. There is no magic to the 10 percent number; the higher the percentage of gasoline that is decontrolled, the lower the price of decontrolled gasoline would be and the lower the tension on the gasoline lines would become. The point is that decontrol of some gasoline supplies would mean less tank topping, shorter gas lines, longer station hours, and so forth.

Incidentally, there is some anecdotal confirmation of the effectiveness of this idea. During the shortage in Boston, the operator of a gasoline station willfully disregarded price control regulations, and sold his gasoline at a substantial premium (up to 50¢ per gallon). The result was that he was open 24 hours a day. Unfortunately, because he violated the price control regulations, he was arrested. Closer to home, last year, when the shortages started loosening up, I noticed that even though I had to stand in line to buy gasoline in my usual neighborhood station, I could buy gasoline whenever I wanted, and with no delay—until midnight—at another significantly more expensive (10 to 15¢ more per gallon) station, also near my home.

The surprising thing to me was how relatively small the differential in cost had to be before some of this "gourmet gasoline" effect came into play. It was particularly surprising in the case of the Boston gas station which was, apparently, the only "gourmet gasoline" station in Boston. It was especially interesting if you believe as I do, that this premium should fall as more stations qualified for decontrolled sales. In fact, I believe that this kind of empirical observation suggests the need for rethinking the use of long-term demand elasticities as a method for analyzing the behavior of prices and demand during short term supply disruptions.

For those who believe in the perfectibility of human institutions, there is the permanent hope that if they were only allowed to make one more change to the allocation system, that they might yet get it right. This was basically the answer proposed by the General Accounting Office in a recent report they published reviewing the performance of the allocation system. But I really wonder who could forecast supplies accurately enough, and who would be wise enough to foresee the effect a contingency might have on the pattern of demand?

Had there been profitable legal ways to transfer some of the relative surplus in the countryside back to the relatively supply short cities, it would have resulted in a much more efficient and equitable allocation of supplies. No inflexible system of allocation can "get it right." We must build in ways for people to trade supplies among themselves in such a way that everyone is better off. Another way, besides gourmet gasoline, in which this might be done is to allow all jobbers and gas stations in areas with relative "excess supply", to trade up to a certain percentage—perhaps 10 percent—of their allocation to areas with relatively low supplies. The problem with allowing trading for profit is that it can be used to subvert the entire allocation system. The problem with not allowing it is that you are left with a system that has no incentive for smoothing out the inevitable disequilibria that will result in any fixed formula allocation system.

THE CHOICES AHEAD

Over the year ahead, the United States will have an opportunity to decide how it will manage any future oil contingencies that it might be confronted with. At present, the legal authority for managing oil shortages resides principally in the Emergency Petroleum Allocation Act (EPAA, P.L. 93-159). This act is the source of the gasoline allocation and price control system, and it also contains the authority for implementing gasoline rationing; it expires on October 1, 1981.

Gasoline Rationing versus Standby Tax with Rebate

Recently, a very healthy public debate has developed regarding the advisability of gasoline rationing for dealing with an emergency, as opposed to that of a standby gasoline tax/rebate plan. There are those who argue against gasoline rationing saying that it would create: a monstrous exceptions bureaucracy; a huge invitation to fraud; a separate currency system rivalling our present one; a bureaucracy as large as the current Department of Energy—20,000 people; and a real risk of having the rationing coupons and their availability not match the demand pattern for gasoline and its availability.

On the other side, there are those who argue against a standby tax, and for gasoline rationing, saying that a standby tax in the event of a substantial shortage would generate huge revenues rivalling the currency now in circulation, perhaps over $400 billion a year, and would cause major problems with the macroeconomy because of the enormous money flows involved, and because the effect on the CPI of higher gasoline prices.

The more you think about these two alternatives, though, the more similarities you discover. For example, the distribution of the coupons is entirely akin to distributing money; therefore, every argument against rationing because of the complexity of discovering who would be eligible for the coupons, and because of the ease of fraud with multiple vehicle registrations, and because of the ease of buying old inoperable cars solely for the benefit of receiving the coupons, is entirely applicable to the rebates of the standby tax, were we to rebate those monies to the same people that would receive the coupons.

Every argument against rationing because of the need for creating a massive exceptions bureaucracy, would also in principle be applicable to the standby tax/rebate proposal, if a large bucreaucracy were created to process exceptions to the rebate system. Also, if product prices were decontrolled, the market price of gasoline under the standby tax should theoretically be exactly equal to the price of gasoline at the station under rationing, plus the white market price of the coupon. The market price under either system is determined, theoretically, by the price level at which demand equals supply. Therefore, the two systems are very similar, and have many of the same problems.

I do not want to delve deeply into this here, but I will make the point that from a distributional point of view, there exists a standby gasoline tax and rebate scheme that can be shown to be exactly equivalent to any given standby gasoline rationing plan. And if people are willing to support rationing because they consider it to be an equitable way to manage a shortage, they should also be willing to support a standby gasoline tax with an appropriate rebate scheme. I especially want to make clear that rationing is per se neither more nor less equitable to the poor than a standby tax/ rebate scheme. It is often argued that gasoline decontrol will allow the rich to get theirs while the poor can not. I would point out that this would also be true of rationing with a white coupon market; that is, the rich would still be able to buy their way out by buying extra coupons from the poor. But, obviously, the poor who had sold some of their coupons must have preferred obtaining that money rather than using that gasoline. In a similar way, the rebates associated with a standby tax/rebate scheme allow us to separately treat the income distribution question from the question of who is willing to pay more for a given volume of gasoline.

There are, however, several operational differences between rationing and a standby tax that argue in favor of the tax, and convincingly so in my opinion. First, a white market for coupons would be a new concept that would take some time to shake out before people understood it as well as the regular monetary economy. The value of the coupons would not be understood as well as the value of money. Second, operation of the coupon system would require a separate coupon economy at the supplier end; this would pose tremendous possibilities for unintended interference with the flow of products, as tickets may not flow with the gasoline as smoothly as might have been intended. Third, because most people would not be as familiar with the rationing coupons as with ordinary money, this would present great opportunities for counterfeiting and fraud. Fourth, the standby tax and rebate would be a lot more explicit in economic terms than rationing and would thus lead, I believe, to a better distribution of the discomfiture caused by a contingency than rationing would. The tax would monetize everything, and it would be difficult to explain to society why one family, for example, should receive three times the rebates that another would. Unfortunately, for that very reason, the tax/rebate system would make for some very difficult political trade-offs; the rationing system, by obfuscating what was really going on, would make it easier politically to make those tradeoffs.

Further, as a very practical matter, a standby/tax rebate scheme would make it possible to avoid some of the difficulties now attached to rationing because of historical circumstances. Rationing is now required to use a particularly difficult system for distributing the ration coupons, i.e., on the basis of vehicle registrations. This is unfortunate because of the difficulty of protecting against cheating in this kind of system, for example, against the possibility of registering a vehicle in several states at one time. This is not a fundamental difficulty of rationing so much as a practical difficulty given the way that rationing as a concept has evolved. In a standby tax/rebate scheme, there is clearly no need for choosing such an arcane method for distributing the rebates.

The Real Issue: Control versus Decontrol

For a variety of reasons I detail in Appendix I, I would expect that a 15 to 20 percent shortage would be a conservative estimate of the magnitude of most emergency oil shortages we are likely to face. If this is a reasonable estimate, I would believe that a standby tax/rebate system would best be able to handle most shortage situations.

However, despite the differences and similarities between the rationing and standby tax options, I believe the real question for the United States to decide is whether or not allocation and price controls will be continued, even if only during an oil contingency. The fundamental question is whether or not the United States will use a market system to do the job of allocating supplies, and use a tax system to absorb windfalls, and redistribute the proceeds, in order to take care of the equity issues involved.

The alternative approach would be to use an allocation and price control system to attempt to physically direct the flow of supplies throughout the economy, through the various levels of suppliers to all consumers, while using a standard of equity which is based on historical flows. Rationing could then be used to channel the windfalls created by the oil contingency back to the general population. The problem with the allocation/price control approach is that it is by its very nature inflexible, and does not allow the trading-for-profit that is the engine of any market economy. The advantage of this approach is that it is superficially equitable and, more importantly, that it protects the entire industrial infrastructure between the oil producer and the oil product customer which might otherwise be damaged severely.

Over the next year there will be a lively debate over the future of oil allocation and price controls. Among the proponents of an extension of these authorities will be quite a large number of refiners, jobbers and gasoline station owners. They will claim that without the sort of protection granted to them under the EPAA they will be left out in the cold in the event of an oil shortage, and will probably go under as a consequence. Others will claim worse damage, that even in the absence of a shortage they will be unable to compete with the integrated companies.

How should we decide these issues? Several questions arise naturally. First, what should be the objective of an allocation/price control system: to guarantee access to all levels of the industry, to supplies priced below market levels at all times, such as the period from 1973 to the present; or to do these functions only at the time of a shortage? If these authorities are triggered, how should they be triggered, for how long, and on whom should what kind of burden fall for extending them? To what extent would these authorities conflict with the goal of promoting an efficient market for oil products? Would there be an alternative way for achieving the objectives of such controls which would be better suited for promoting efficient markets and still satisfactorily address the various equity issues involved?

At this point, I should review the three broad objectives that I originally proposed might be addressed by a strategy to manage oil contingencies: first, a market efficiency objective; second, an equity to the general population objective; and third, an objective of giving some protection to the market structure of the industry in the event of a contingency.

There is no question that decontrol with some kind of tax and rebate would best serve the interests of efficient markets. I also believe that a rebate provision could be designed to best address the equity issues involved with the general population. But to pretend that decontrol with a standby tax and rebate would solve all the political problems of contingency planning, would be to ignore the political reality of the industrial infrastructure that would be damaged by an oil shortage—the independent gasoline and heating oil dealers, the jobbers, and the independent refiners, among others. The answer to this problem in 1973 in the EPAA was simple: in times of emergency, you protect all of the elements up and down the chain, all the way down to the retail outlet; and this protection would last until it was removed by explicit act of the President, and subject to disapproval by either House of Congress. This is why the United States still has gasoline price and allocation controls almost seven years after the original emergency occurred. And this is also why we would probably be forced to reimpose them if another emergency occurred.

A Suggestion

I would make the following suggestion as one way of avoiding the reimposition of full battle regalia allocation and price controls in the event of an emergency, while still addressing the previously mentioned objectives. As I noted, we can deal with the question of equity to the general consumer through a satisfactory rebate provision; the next question is how to deal "equitably" with the actors in the distribution chain. In theory, it is possible to address separately the market efficiency and equity objectives involved in any allocation system. For example, allocation and price controls could be triggered into effect during an emergency, for a limited period, perhaps to be extended by the President for an additional short period—subject to Congressional affirmative vote—but then to be extended only by specific Act of Congress. This would protect the actors in the distribution chain. However, in order to promote market efficiency, we could allow those at the bottom of the distribution chain who received the price controlled allocations, to resell the physical products in a decontrolled market, perhaps to other dealers, or to their own suppliers. I call this a "marketable allocation." Because we would have dealt with the consumer equity issues in the structuring of a rebate of the tax, we could decontrol without fear, while satisfying the other objective of protecting the distribution chain.

This kind of proposal would meet the market efficiency objective, the consumer equity objective, and the objective of giving some limited protection to the distribution chain, while allowing products to flow to where they are most valued.

I do not want to imply this suggestion has no problems beyond those I have already discussed. For example, whatever protection is granted to this sector will tend to encourage irresponsible behavior on its part. If these actors know that in the event of an emergency they will be protected, they will tend to disregard taking expensive steps in the present in order to protect themselves against a future possible contingency.

Other Problems with a Gasoline Standby Tax/Rebate Proposal

In fact, there are many problems with the standby tax/rebate proposal. For example, it would seem best to place the windfall tax on crude oil, rather than gasoline. Crude oil, after all, will be the item in short supply in the event of a contingency; at such times there will be a surfeit of refineries relative to the supplies of crude oil. When a shortage occurs, the price of crude oil will go up to the level where the demand for crude oil equals its supply. To tax the windfalls only on gasoline would be to allow windfalls on other products such as heating oil and jet fuel, among others. Also, a tax on gasoline would not produce revenues large enough to deal adequately with the equity issues involved. Worse yet, taxing one product, but not all others, would cause unforeseeable, and perhaps bizarre market distortions; for example, we could end up with artificially worsened gasoline shortages, at the same time that gluts were created in the heating oil and jet fuel markets.

But there are also a host of other issues. For example, we would have to be careful how a standby tax on crude oil would be triggered lest we discourage the storage of oil prior to an emergency. If we taxed away windfalls on oil stored prior to an emergency, this would have the undesirable effect of reducing the present incentive to stockpile crude oil; doing this would simply eliminate the socially desirable windfall to those who had been farsighted enough to protect themselves. If would seem that the tax should go only on oil produced after the contingency occurs. If so, it will be necessary to be careful as to how such a tax would phase out after the contingency was resolved. If the tax disappears immediately, it would tend to encourage the delay of production till after the contingency ended in hopes of receiving higher prices without a tax. If the tax were permanent, it would have other effects such as discouraging the search for new oil. These are not simple questions that have simple answers, but they will have to be studied and answered if such a system is to be adopted.

The Strategic Petroleum Reserve

I also should discuss the Strategic Petroleum Reserve (SPRO). We must proceed to fill it, immediately, and the faster the better at this time. Fortunately, the US Congress has now legislated that its fill be recommenced. In the event of an emergency it will pay for itself many times over. What isn't so clear is what decision rule should be used to release oil from the SPRO, and what price it should be released at. If it is released at anything below the market price, it will be discharged at its maximum rate with the consequence of simply replacing imports barrel for barrel. This would not be a good use of the SPRO. If, on the other hand, it is released at high spot price levels, the public will be encouraged by refiner interests to believe the government is price gouging the consumer, rather than simply charging the affected refineries what the oil is worth.

My own guess is that we will have serious problems in determining how best to release the SPRO oil; that there will be a tendency by the Government to hoard it; and that when the decision to release the oil is made, there will be a tendency to release it at lower than market prices. I think these are all serious problems, but they are almost academic at this point given the 92 million barrels we now have in the SPRO.

CONCLUSION

To sum up, I am gratified that there is finally starting to be an awareness in the United States that energy contingency planning is of the utmost importance for our economy and our political system. I am, however, troubled that so much of the activity seems to miss the point that use of market mechanisms could be invaluable for managing a shortage in an effective and least costly way, if equity and income transfer considerations were explicitly considered and built into the societal response prior to the occurrence of the contingency. More contingency managers in the government, to answer ever-increasing requests for exceptions relief to an inflexible allocation system, may be an inappropriate strategic response to the problem.

The bad news is that the United States is ill-prepared to handle an oil contingency at the present time. Its allocation/price control system for gasoline is defective in the event of an emergency, and I do not think that any number of fine adjustments will be sufficient to substitute for market-like flexibility which very deliberately could be built into the system itself. Standby gasoline rationing is no panacea either, especially when considering the time it will take to implement. And if things were bad in 1979, they will be much worse if standby rationing is triggered while allocation and price controls are still in force.

Prior to October 1, 1981, the United States should consider implementing market mechanisms to help handle dislocations. After that date, the United States should consider enacting a standby crude oil tax/rebate plan to be triggered during emergency oil shortages. If Congress wishes to provide some form of relief to protect the distribution chain in the event of an emergency, it should consider allowing the trading of oil products among dealers during a shortage —this is the so-called "marketable allocation" option I discussed previously.

The good news is that the United States is not as far from this standby tax/rebate system as we might imagine. The United States has a windfall profits tax on domestically produced crude oil; it provides the basis for the enactment of a standby emergency tax. Appropriate structuring of rebates would be another question, though. However, these difficulties should not be allowed to stand in the way of developing an explicit strategy for managing emergency oil shortages in as efficient and equitable a way as possible.

APPENDIX I

What is the likely magnitude of the oil
shortage and does that affect the
appropriateness of the response?

Because the market clearing price for gasoline would be a function
of the size of the shortage, one should ask whether either system
(gasoline rationing vs. tax/rebate) should be preferred on the basis
of this criterion. To the extent that the standby tax/rebate alternative involves the transfer of progressively larger amounts of money,
and to the extent that this makes it progressively more difficult
to achieve a socially acceptable standard of equity, it would seem
that the larger the shortage, the less attractive the standby tax/
rebate alternative would become. It would also follow that the
more relatively attractive the use of non-market mechanisms such
as rationing with price controls would become.

The next logical question is how large a contingency are we
likely to face? There are at least two different ways of estimating
what the loss of US oil imports will be, based on a given interdiction
of oil supplies in the world oil market. Under one scenario, the
United States may only have its oil imports reduced as much as any
other country, i.e., all countries will lose an equal share of their
oil imports based on the net percentage reduction of oil in world
trade. For example, if the United States, importing 8 million barrels
a day (MMBD) out of 30 MMBD in world trade, say, the United
States on this basis would be expected to lose 8/30 or 27 per cent
of any net reduction of oil in world trade.

In the alternative, each country could have its total oil consumption, rather than just its oil imports, reduced by an equal percentage.
This, of course, would reduce US oil imports by more than the first
alternative. For example, if the United States consumes 18 MMBD
out of the 50 MMBD consumed in the free world, the United States
might be expected to lose 36 per cent of the net reduction in oil
available to the world.

Let us further speculate that Saudi Arabian oil production was
interdicted for some reason, that the world thus lost 9.5 MMBD of
production, and that this was offset partially by increased production
from previously spare capacity of about 4 MMBD. The net shortage
would thus be about 5.5 MMBD. Under the first alternative, it
would then follow that the United States would lose about 1.5 MMBD
of its oil imports (note the United States now only gets about

1.3 MMBD directly from Saudi Arabia, even so, the loss would be bigger than just that oil obtained directly). Under the second alternative, the United States would be expected to lose about 2 MMBD.

Now, it is unlikely that Saudi Arabian production would go entirely out—except in the event of something beyond the country's control. If, instead, the Saudis attempted another US oil embargo, they would in all probability not cut their production below 3 to 5 MMBD; in such an event, the net shortage on the world oil market would be less than 2.5 MMBD. Even if a few other oil producing countries sympathetically curtailed some of their oil production, the total shortage may not exceed 5.5 MMBD even though, in theory, it could be much greater. During the 1973 Arab Oil Embargo, production cutbacks were not undertaken in many countries that might have been expected to support the Embargo. In fact, countries such as Iraq actually increased their production.

Based on these arguments, one would expect a relatively manageable shortage in the United States. On the other hand, the oil market is today less flexible than it was in 1973-1974 when the US Arab oil embargo was a failure, as an embargo of the United States. I call it a failure as an embargo of the United States because the shortage was really felt equally everywhere. With the active assistance of other oil producing countries, an embargo of the United States could conceivably involve US shortages greater than 2 MMBD.

Thus, one would imagine that a US shortage of 2 to 3 MMBD would require a very large world oil interdiction. I would think it unlikely under most circumstances that the United States would have to face much more than that. If this is correct, it can help put the US oil contingency management problem into some context. We should find methods that could help us handle reductions of up to 15 to 20 percent of our oil consumption. If we also arranged to draw down the Strategic Petroleum Reserve, and make maximum use of other alternative fuels—such as gas, coal, and nuclear based electric energy—a 20 per cent shortage would appear to be a conservative estimate of the magnitude of most emergency oil shortages. There are situations that could easily be conjectured that would involve still larger shortages, but discussion of methods for treating these are outside of the scope of this paper.

Finally, I should also note that the shortage is usually aggravated at the beginning of an oil contingency by the fact that consumers, refiners and everyone else begin to accumulate larger stockpiles for speculative purposes. Crude and product decontrol should significantly reduce incentives for speculative buildup.

APPENDIX II

A Gasoline Tax as an Emergency Measure as Opposed to a Measure for Promoting Long-term Demand Restraint

This paper deals only with the issue of using a tax (preferably on crude oil) only to capture the windfalls that might be created in an emergency oil shortage. It does not deal with the pros and cons of enacting a long term tax on crude oil or gasoline to promote long-term demand restraint.

There is an important distinction to be drawn between these two concepts. The first is simply intended to help address the equity question in the event of an emergency oil shortage, so that decontrol of the oil products market can be allowed in order to facilitate an efficient market allocation of those products. The second concept, that of taxing the product now in order to help restrain future demand, depends on other complicated arguments. While both concepts may each have some validity in their own context, it is disingenuous and a disservice to the public debate to blur those distinctions.

APPENDIX III

Clarification on Allocation and Price Controls

It is not well understood that, for most practical purposes, allocation controls go along with price controls, and vice versa. Having one without the other does not amount to much. To see this, consider the imposition of allocation controls without price controls, where company A could be forced to supply company B, if B desired. This would mean that in order to meet the requirements of allocation controls, company A would only have to offer oil supplies to company B, but it could do so at a price that would clearly make it unattractive to B. Thus, this requirement is not meaningful. Viewed differently, the only way the concept of allocation makes sense, is when the government is forcing the allocation of a valuable good—for example, when it is forcing the allocation of oil supplies which are price controlled at below market levels.

Similarly, if price controls are imposed without allocation controls, they will not address one of the principal reasons for their imposition. We are not here referring to issues of long-run supply or long-run demand which are raised by price controls. Instead, we refer to the fact that price controls without allocation controls do not guarantee that company A will make the underpriced product available to company B. For example, company A may choose to retain all or most of its supplies of price-controlled products for direct sale to final consumers. This means that price controls without allocations would lower prices to final consumers in the near term, but would not protect the general distribution chain in the event of a contingency.

For these reasons, allocation and price controls are usually discussed together.

Chapter 20

Institutions for Managing Energy Shortages

Fritz Lücke*

I should like in this short paper to develop some basic ideas on an energy policy as reflected in the policy conception of the Federal Ministry of Economics and to put this into perspective with alternative policy approaches.

Let me start with briefly delineating the energy situation of the Federal Republic of Germany. On the one hand it is, as many other countries, depending on a reliable and equitably priced energy supply in order to maintain its competitiveness as an exporter to the world markets, its major source of income. On the other hand it is a country of very limited indigenous energy resources: almost 100 percent of its uranium, 96 percent of its crude oil and 60 percent of its natural gas is imported. The very exception is coal, but its 26 percent contribution to our TPE is subsidized with some 6 billion Deutschemarks per annum.

Germany's overall import dependency in 1979 was about 57 percent of total TPE with the consequence that Germany not only heavily depends on the smooth functioning of the world energy markets, but also that all short or long-term risks have an immediate

*Federal Ministry of Economics, Bonn.

225

impact on the German energy supply situation and hence on her economy.

Two qualifications therefore characterize the energy policy conceived against this background. It is necessarily part of the general economic policy—it is essentially a longer term policy. Let me elaborate on these two qualifications.

If the general economic policy objective is to further economic growth and public welfare by active competitive participation in the world markets on the basis of a free market economy, the prime function of a complementary energy policy must be defined as providing conditions and orientations within which a sufficient, reliable energy supply at fair prices can be safeguarded so as not to hamper economic growth. Adaptations to structural changes on the supply or demand side thus remain the task of the market participants, and it is only to the extent that the market forces do not or not sufficiently react that government assistance or intervention is called upon.

It is therefore an essential element of German energy policy not to interfere with energy prices, acknowledging that rising prices are the most efficient incentive for energy conservation, substitution and for the development of new resources and new technologies. To a certain extent this has already materialized since 1973, and the list of new projects which have become economic or do now appear within reach in industrialized consuming countries is already quite impressive. This in spite of the fact of long lead times, price controls in a great number of countries, and last but not least much lower price increases in real terms. But everybody agrees that this is only a start, that much more must be done, and the question may in fact be asked which policy best serves our long-term objectives: the one which attempts to dampen international and national price developments whenever possible, or the one which accepts rising energy prices and in particular the prevailing price-setting forces as an element in its long-term strategy.

Of course, this can hardly apply to such erratic and wanton price explosions as the one we have just had with a rise of over 140 percent, and some of us may recall with some nostalgia the times when consumer governments and oil companies were fighting against the indexation of oil prices.

This elementary role of the energy market price does also not imply that over and above the policy orientations there is no need for supplementary government action, be it to help the needy, be it to accelerate the adaptation process, be it to bridge transitory phases. That is for instance the case in Germany with certain subsidies to low income groups, with a large and comprehensive conservation program and with the maintenance of the German coal industry at very high costs in the past. And the IEA is certainly right in advocating further strengthening of efforts from all its Participating Countries. But a very clear distinction is made in Germany between the responsibilities of the market participants and governments, and it is for this reason that hesitations have been expressed against defining common stock policies beyond the compulsory stocks-commitment or against so called cool-down-procedures in times of overheated oil markets, and it is therefore that regulatory measures like the famous speed limits are considered as the last resort.

I said at the beginning that energy policy in my view is also essentially a longer term policy, and this of course is fairly obvious given the structural and social changes, the lead-times, the investment involved. This raises the complicated issue of the antagonism between short-term politics and longer-term targets, as they have been discussed in various *fora* do not appear to be of too much help in this context. Over and above a functioning price mechanism there is not very much that could be done in the short run to substantially reduce oil imports or consumption before you reach the rationing phase.

However energy policy in our days is no longer feasible in splendid isolation. The political, social and economic interdependence of our economies makes coordination and harmonisation indispensable, and this applies in particular to an energy economy based on the delineated policy approach. Cooperation, in particular in the IEA, has made a promising start. We expect that the qualitative strategy developed by IEA-Ministers last May of pinpointing areas in participating countries, where additional efforts are warranted and to subject these efforts to a strict review procedure, will bring us a great step forward.

This is however only one side of the necessary coordination. It has become clear by now that coordination and cooperation with the countries of the Third World, and not only the producers, but the non-oil-producing consuming countries, is essential for the solution of our long-term energy problems. This is one of the messages which the Venice summit has clearly stated.

In short, an energy policy based on the free functioning of the world markets is only successful against the background of a broad international understanding between all parties involved, industrialized consuming countries, oil producers and developing consuming countries, and this task has yet to be tackled.

This seemingly smooth scenario does of course not deny the risk of short-term market disruptions. And there we are back at the more technical institutions for managing energy shortages. We have the oil-stock-piling corporation, formed by the German oil industry and trade to take care of the 90 day stock-piling commitment on a private basis and through private financing—another example for the basic policy approach outlined above.

We have the IEA emergency system for supply shortages of 7 percent or above, supplemented by a set of national contingency plans and regulations. And the IEA has developed a whole system of yardsticks and ceilings, an adjustment system for these ceilings, a quarterly monitoring and a procedure for consultations on stock policies in order to deal with market disruptions below the 7 percent trigger threshold.

It is our conviction that with this arsenal we should be appropriately equipped to cope with the short-term risks and that our attention and efforts should now be focused on the medium- and longer-term progress necessary to achieve the required structural changes and thus to secure our energy supply for the future.

Chapter 21

State Oil Trading and the
Perspective of Shortage

Øystein Noreng *

INTRODUCTION

Historically, few sectors of the international economy have changed as profoundly in such a short time as has been the case with the international oil market in the decade from 1970 to 1980.[1] The abrupt rises in the relative price of oil in 1973-4 and 1979-80 are the most apparent aspect of this change. However, the price rises often seem to overshadow more profound, structural changes that have taken place. The nationalization of oil extraction in most of the oil exporting countries, followed up more recently by an increasing nationalization of oil trade, have transformed the institutions of the marketplace and the organization of the international oil trade. In a dynamic perspective, these changes are likely to be of greater importance than occasional price increases, which later can be at least partly undone by inflation, as has also been demonstrated in the decade from 1970 to 1980. Reasonably these structural changes are irreversible; neither nationalization of oil extraction nor of oil trade are likely to be undone; consequently the changes in the world oil market from 1970 to 1980 are not merely the expression of economic fluctuations, but part of a historical process. This historical process concerns the bargaining relationship

---

*Assistant Professor, Oslo Institute of Business Administration.

between oil exporting countries and oil importers, both companies and governments, and it concerns the question of power in the oil market, as the ability of either side to pursue its interests and to control important events.[2] Thus, the process of change in the international oil market concerns the political economy of international oil, giving new premises for the behaviour of the market.

In general, the process of change indicates a transfer of power from the oil importers to the oil exporters, and the latter have during the 1970s become much more able to pursue their interests. Also, their potential for controlling important events in the international oil market has increased. However, this potential has not always been fully utilized. Indeed, during the 1970s the oil exporting countries have shown a remarkable mixture of assertiveness and reluctance, sometimes acting to benefit from a favourable situation in order to implement desirable changes, sometimes appearing extremely cautious in light of a situation that could be used more fully to their advantage. This contradictory behaviour on one hand indicates disagreement and tension among oil exporting countries, on the other hand it indicates uncertainty and a lack of strategy in an entirely new situation. Thus, the process of change in the world oil market is a contradictory one. In this perspective it is legitimate to raise the question whether the oil exporting countries are prepared to use the power given to them by the new market situation, and even whether they really want this power.

From the point of view of pursuing the interests of the oil exporting countries, the decisions to nationalize oil extraction, and more recently oil trading, were rational, as they gave control of important events affecting their own basic industry. However, these moves also vest the oil exporters with some delicate decisions and awkward dilemmas that are of an international dimension. Even if in normal circumstances only a few of the oil exporters have a marginal function in the market, the precarious balance between supply and demand, particularly in a long-term perspective, means that for most of the oil exporters, control of their own oil industry means directly some degree of control of the world oil market, and indirectly some degree of influence with the world economy, at least in a dynamic perspective. Thus the oil exporters have constantly to strike a balance between their own interests and the perceived interests of the world community, between short-term concerns for the world economy and long-term concerns for the world's energy supplies. Within this context, the nationalization of oil trade also means that the oil exporting countries have to strike the balance between the needs of different clients, between developing countries and industrial countries, and also between different developing countries and different industrial countries. Clearly, this situation gives considerable political leverage to the oil exporting countries, the counterpart of which is exposure to political pressure. Consequently, relations between oil exporters and oil importers are becoming more complex, meaning that oil is less and less a conventional commercial commodity, and more and more a political commodity, with oil trade being closely related to power politics. This politicization of oil trade may not be entirely in the interest of the oil exporting countries, as it to some extent could limit their commercial and economic freedom of action. As political considerations impose themselves upon the marketing of oil, the maximization of income becomes more complicated. Furthermore, the increasing degree of state trading in oil contributes to the politicization of the oil market, by bringing the oil exporting governments more directly into the transactions. Whereas it would be erroneous to describe the historical role of the international oil companies as "non-political", they did provide a buffer between consumers and oil exporting countries. The increasing degree of state trading in oil to a large extent does away with this buffer, exposing the oil exporting governments more directly to the demands of the consumers. In normal circumstances, and with a fairly slack world oil market, this may be no acute problem. In a situation where international oil demand reaches a level which the oil exporters are unwilling or unable to fulfill, there is likely to be a crunch for the new structure in the world oil market, as the oil exporters will have to allocate scarcities. Thus, together with asserting their interests the oil exporters have also taken over some awkward tasks and some delicate responsibilities.

THE INSTITUTIONAL CHANGE IN THE WORLD OIL MARKET

The nationalization of the oil industry by the oil exporting countries and their subsequent take-over of oil export marketing introduces a historically new structure in the world oil market. Paradoxically, the nationalization of oil extraction and oil marketing makes the world oil market appear somewhat more like a free market, in the classical sense, than was the case previously. Traditionally, the world oil market has not been characterized by the play of free market forces in the sense of neo-classical economics, implying mutually independent buyers and sellers and the absence of monopoly or oligopoly, or monopsony or oligopsony. On the contrary, most of the oil traded internationally in this century, at least until the late 1960s, has been the subject of transfers within vertically integrated international oil companies.[3] Only a minor part of international oil trade involved open commercial transactions between mutually independent buyers and sellers.[4] Through this integrated pattern of organization, the dynamics of oil demand and oil supply were largely synchronized, with decisions concerning extraction, refining and marketing being essentially made within the same organization. This reduced the potential for gluts and shortages.

With the nationalization of oil extraction and trading, the vertically integrated pattern of organization has been broken up, decoupling the dynamics of oil supply from those of oil demand. Consequently, the potential for gluts and shortages has been increased. Nationalization of oil trading reduces the role of the international oil companies in distributing oil, or oil scarcities, with this function being transferred to the oil exporting governments and their national oil companies. An essential element of the new market structure is that international oil trade now to a large and increasing extent takes place between mutually independent buyers and sellers. Apparently, this could mean that market forces in a classical sense would be more important than previously in determining patterns of oil demand and supplies, and oil prices. However, given that supplies of oil to the world market are dependent upon a fairly small number of oil exporting countries, which, as pointed out, increasingly market the oil themselves, whereas the buyers in the world oil market now are several hundred companies and governments, the situation is highly asymmetrical. Two basic differences between the demand and the supply side can be pointed out; the relevant level of decision-making is governments on the supply side, companies essentially on the demand side; demand appears to be much more spontaneous than supplies of oil. Given that some of the oil exporting governments enjoy a considerable freedom of action in oil policy and economic policy, the level of oil supplies entering the world market to some extent is arbitrary; indeed the balance between oil demand and oil supplies appears to be essentially politically determined, depending upon the preferences of a few oil exporting governments; thus, the price of oil also appears to be fairly arbitrary, and to a large extent determined by political circumstances.

The historical pattern of organization of the world oil market, with oil trade being controlled by international oil companies, reflected two realities; that the oil exporting countries were dominated politically by consumer interests, and that the world oil market lends itself to regulation. The conditions of the historical concessionary system in practice implied a transfer of sovereignty from the oil exporting countries to the international oil companies of matters relating to oil.[5] The background was the colonial or the semi-colonial status of the bulk of the oil exporting countries at the time oil was discovered. The result was that depletion decisions were taken essentially with the companies' interests in mind, not those of the host countries, as is demonstrated by the declining reserve-to-extraction ration.[6] This also meant that the consumers got signals from the world oil market that were historically finite; increasing quantities of cheap oil were dependent upon a situation of political dominance that could not last indefinitely. In hindsight it could be argued that the oil importing countries today might have been better off had they not received these historically finite signals during the 1950s and 1960s. Regulation, through private monopolies or oligopolies, or through government, is practically as old as the oil industry itself.[7] Since the very beginning the oil industry has had a propensity for vertical integration, for oligopoly, and for regulation of the market rather than competition. The reason is essentially that energy markets because of low price elasticities of supply and demand easily lend themselves to centralised administration. Thus, the price mechanism alone is no good regulator of supply and demand in the energy markets, and particularly in the oil market, given the relative indifference to supply and demand to price variations, at least within ranges known historically. Instead the price determines the distribution of income related to oil, i.e. the distribution of the oil rent (the sum of profits and income related to oil) as well as its size. This background for persistent attempts by the oil industry to regulate its environment and by governments to regulate the energy markets.

From the industrial consumers' point of view the regulation exercised by the international oil companies over the world oil market was efficient and desirable throughout the 1950s and 1960s. The low price elasticity for oil demand indicates that consumers in general give a higher priority to stable and secure oil supplies than to pricing; this is confirmed by consumer behaviour in periods of crisis. In this perspective, it could be in the consumer interest that the regulation of the world oil market be exercised by the oil exporters, as the international oil companies no longer have the power to perform this role. From the consumers' point of view, price instability and potential supply shortfalls are less desirable than a stable market, even if this would mean the entrenched power of the oil exporters, and gradually rising oil prices. In this perspective, for the consumers, a weak OPEC could be less desirable and more fearful than a strong

OPEC.[8] Thus, for the consumers, the erosion of OPEC's control with the world oil market, as occurred in 1979 because of high demand and the Iranian crisis, was regrettable. Oil exporting countries benefited from rising prices, and paradoxically, from their own cartel losing control of the situation. Intermediaries, i.e. traders, brokers and oil companies, benefited from rising prices and chaos in the market, giving substantial differentials and profit margins. Thus, the events of 1979 have demonstrated that OPEC is not directed against consumer interests, and that for the consumers, a regulation of the world oil market by OPEC is preferable to no regulation. Oil prices have hardly ever been determined by classical market forces, and there is little reason to assume that market forces will be more effective in setting optimal prices for oil in the future than they were in the past. Instead, pricing patterns could easily reflect relative influences of particular interests, and be partly irrational in addition to being arbitrary. As argued earlier, oil prices are to a large extent arbitrary, and for the consumers it is preferable that the arbiters be an organization of governments rather than particular interests with no identification of political responsibility. It is evident that OPEC's aim is not short-term maximization of profits, but that it is pursuing a complex set of objectives, one of which is the long-term aim of keeping oil prices in line with the cost of alternative sources of energy.[9] Thus, the anger that has been expressed in many industrial oil importing countries over OPEC is essentially misdirected, ignoring the fact that OPEC's regulation of the world oil market often is a moderation of oil prices.

The nationalization of oil export marketing constitutes the second phase of the institutional and organizational change in the world oil market. Direct sales by oil exporting countries have increased significantly in the 1970s; in 1973 they accounted for about 8 per cent of international oil transactions, measured by volume; by 1979 they represented about 42 per cent of international oil trade; by 1980 more than half the oil entering the world market is likely to be sold directly by the oil exporting countries and their national oil companies.[10] There has been a remarkable increase in both state-to-state deals, whose share of world oil trade has increased from 5 per cent in 1973 to 17 per cent in 1979, and commercial sales by oil exporters, whose share has risen from less than 3 per cent in 1973 to almost 26 per cent in 1979. However, volumes sold through state-to-state deals have risen more regularly, indicating that the surge in state commercial sales may reflect a take-over of previous private contracts, and that the development of state-to-state deals is lagging behind volumes available to the oil exporters for direct sales. In the OPEC countries, direct national sales of oil represented more than half the export volumes already by 1978. In practically all OPEC countries there is a clear preference for direct state marketing of oil. This includes Saudi-Arabia, which also on this point has been fairly conservative, where nevertheless the share of the national oil company Petromin has been increased at the expense of Aramco. Also, in oil exporting countries outside OPEC there is a preference for state trading in oil. All Mexican oil exports are handled by the national oil company Pemex; in the United Kingdom the national oil company BNOC disposes of half the oil extracted; in Norway the share of the national oil company Statoil is approaching half the volume extracted.

In recent years OPEC has been defining a new trading policy as part of the overall strategy elaboration. The basic guidelines have been discussed by one of the members of the strategy group, and they appear to be the following:[11]

— crude oil will be sold by the national oil companies of the OPEC countries to foreign refiners on long-term contracts;
— contracts will have a destination clause forbidding the resale of oil by the refiner without specific permission by the supplier;
— the lifting of crude oil is to take place regularly throughout the year with little tolerance for liftings above or below contract volumes;
— financial terms will be stricter as to credits and methods of payment;
— there will be some kind of preferential treatment for the national oil companies of the oil exporting countries, usually by more indirect methods than by a price discount;
— the role of intermediaries, such as traders and brokers, will be considerably reduced, though not probably entirely eliminated; this will also reduce the role of the spot-market;
— there will be a close supervision of the profits made on oil by refiners and marketers;
— there will be an increasing effort not to sell oil alone, but only in connection with products, i.e. progressively introducing "package contracts";
— there is likely to be some preferential treatment of national oil companies of industrial countries, though this is less certain and more subject to local variations than other aspects of the new policy.

SIGNIFICANCE OF THE CHANGE

The guidelines of the new trading policy should be seen as long-term goals; there is neither a political will nor an organizational capacity to implement the changes in trading immediately. Even if there seems to be agreement in principle over the new trading policy, some countries are less likely than others to give a high priority to the change. This especially concerns Saudi-Arabia, where it will be difficult to implement the new trading policy as long as Aramco is the main outlet, keeping close organizational ties with a number of international oil companies. Thus, as long as the new trading policy is not fully implemented by the major oil exporting country, its importance is fairly relative. However, the new trading policy could be seen as realistic, anticipating certain changes likely to take place in oil marketing practices; such changes would represent a continuity of "oil nationalism" and of OPEC policy. In the following, the discussion will assume that the new oil trading policy will be fairly widely implemented during the first part of the 1980s.

The economic significance of the new trading policy is less to directly maximize income for the oil exporting countries than to increase their share of the oil rent (i.e. the sum of profits and income to be made from oil), and to stabilize the market under the control of the oil exporting countries. Limiting the right to resale and imposing a regularity of liftings can be seen as a way to reduce the volumes of oil reaching the spot market, as well as to reduce seasonal or occasional price fluctuations. This gives the OPEC countries a greater regularity and stability of income. The stricter financial terms also work in this direction. Furthermore, the new trading policy transfers from the exporter to the refiner and marketer the burden of adjusting to seasonal demand variations. Thus, the refiner and marketer will increasingly face the choice of either keeping large inventories, at corresponding cost, or not being able to cope with demand. The limitation of the right to resale works in the same direction. This could either lead to a profits squeeze for the refiners and marketers, or to the price going up for the consumers, or both. With reduced flexibility in the intermediary market, consumers will increasingly have to address themselves to the oil exporting countries in case of needed additional supplies, e.g. because of weather conditions. This, of course, gives the oil exporting countries substantial leverage with the oil importers, so that marginal increases in oil supplies might be traded for price increases or political concessions.

Thus, the political significance of the new trading policy is to increase the direct bargaining power of the oil exporting countries in relation to the oil importing ones. The reduced importance of traders and brokers together with a potential profits squeeze in refining and marketing also weakens the overall bargaining position of the oil importers. This is likely to facilitate the OPEC policy of increasingly selling refined and petrochemical products together with crude oil. The potential profits squeeze also weakens the overall bargaining position of the oil companies in relation to the oil exporting countries.

The purpose of restricting spot sales, and the right of refiners to resell oil, is to limit what is often seen as unjustifiable high profits among dealers and brokers.[12] This indicates a preoccupation with the entire cost structure of oil and oil products for consumers, which may be reasonable from a strategic point of view, and which fits with the goal of gradually adjusting oil prices to the cost of substitutes. However, the spot market may be reduced in importance, but it is unlikely to dry up entirely, partly for reasons of control, partly because some governments and national oil companies in moments of tight demand are likely to sell oil in the spot market.[13] Thus, the emerging trading pattern in the world oil market gives an impression of increasing differentiation, even a dichotomy between long-term oil contracts as part of state-to-state deals and short-term or spot transactions. Generally, the oil exporting countries tend to give a high priority to honouring oil supply contracts concluded under state-to-state deals, whereas short-term or spot transactions by their very nature do not give any security of supply beyond the time horizon for particular deals. Even if the new oil contracts give the seller full discretion in establishing the price of oil, retroactively to and increasing, there has historically been a tendency for oil traded under state-to-state contracts to be more moderately priced than oil traded in short-term or spot transactions in periods of tight demand, and it would be surprising if this would not also be the case in the future. Keeping oil prices of state-to-state deals below spot market prices means that the oil exporting countries renounce on income for the benefit of oil importing countries or consumers in periods of tight demand, even if this can partly be compensated for by keeping oil prices of state-to-state deals above those of spot prices in periods of slack demand. Paradoxically, this could also mean that fairly secure long-term oil supplies could be less expensive than insecure short-term ones. Apparently, this contradicts the theory that secure oil supplies have a political rent, and therefore might have a higher price.[14] However, this could confirm another theory, that governments, also the oil exporting ones, are not short-term profit maximizers, but rather operate with a complex set of goals and a fairly long time horizon; in this perspective the political rent attached to secure oil supplies could be taken out in other forms than money, e.g. in relation to goals of foreign policy, trade policy or industrial policy.[15]

The emerging oil trading pattern points to a highly fragmented market, characterized by a multitude of different deals, with diverging prices, supply conditions, payments conditions. This means that customers are not treated equally by the suppliers, and that the suppliers do not have identical objectives in relation to their customers. It also confirms that short-term income maximization in many cases is not the sole, or even the dominant objective of the oil exporters. Thus, the emerging trading pattern hardly conforms to the criteria of an open, competitive market. This underlines once more that oil is a "political" commodity, whose value to buyers and sellers is not fully reflected in the price alone. The emerging oil trading pattern gives a high degree of selectivity to the suppliers, giving the oil exporting governments an ability to use oil supplies as an instrument to pursue policy goals in other fields of interest. This can be described as "politics through trade". On the other hand, there is what can be described as "trade through politics" with oil importing governments willing to give political favours in return for secure oil supplies, or some kind of preferential treatment in relation to oil. Thus, the new trading pattern implies a further politicization of the world oil market, with oil increasingly being linked to other issues, and whose time horizon may go beyond that of the oil transactions. In this way, the "new" oil market is politically more flexible than the "old"one, and this is a rational evolution to the extent that income maximization alone is not the sole or dominant preoccupation of the oil exporting governments. On the other hand, the emerging trading pattern gives an impression of commercial rigidity which makes it legitimate to question its economic flexibility and rationality, particularly in times of perceived unsatisfied demand.

STATE OIL TRADING AND PERCEIVED UNSATISFIED DEMAND

The emergence of a strong spot market during periods of crisis is an important aspect of the breakdown of the historical institutional order in the world oil market. Twice during the past decade, in 1973–74 and in 1979–80, a large proportion of oil traded internationally was the subject of short-term or "spot" transactions, with prices rising substantially above the f.o.b. prices of the oil exporting countries. For example, in the latter part of 1979 the spot market may have accounted for close to 25 per cent of international oil trade, with prices 25 to 100 per cent above the f.o.b. prices of the oil exporting countries. By comparison, in earlier crises, as in 1956 and 1967, the integrated pattern of organization based on the international oil companies managed to shift output fairly easily from one source to another, avoiding dramatic price increases. In the 1970s, by contrast, the established order was breaking down, and old trading patterns were unable to cope with the situation, as the dynamics of supply were largely decoupled and desynchronized from those of demand.

In an economic perspective the surge of the spot market is the result of a real or perceived unsatisfied demand for oil, as during the OAPEC output cutback in 1973–74 and during the Iranian crisis of 1979. Because of the low price elasticity of oil demand, a cutback of a few per cent of regular supplies can lead to major price rises, as this is necessary to clear the market and bring about a real or perceived balance between demand and supply. So far, after periods of real or perceived crisis, spot market prices have fallen again, but not their previous levels, and they have also triggered major increases in the f.o.b. oil prices of the oil exporting countries. Usually, a perception of unsatisfied demand creates a heavier competition for oil, with supply considerations being more important to consumers than price considerations, explaining the emergence of an upper layer of the market for customers who cannot be supplied otherwise.

In the perspective of political economy the rise of the spot market is related to the size and distribution of the oil rent and the question of control of the world oil market. The nationalization of oil production and oil trading together with price controls on oil products in many oil importing countries have reduced the basis for profits of the international oil companies, and they have to some extent been short-circuited by state-to-state deals. Consequently, their share of the international oil market has diminished and their relative part of the oil rent has decreased. The surge in the spot market, as occurred in 1979, and the rise in spot prices as opposed to official f.o.b. oil prices mean that the position of intermediaries improves. There is a fairly systematic trend for product prices in periods of crises to align themselves with spot products prices, which in such situations invariably are much higher than average prices, thus increasing the refiners' margins. Also, the chaotic situation of crude oil prices in periods of crisis means that large differentials and margins exist in the market. Thus, in 1979 not only did spot crude oil prices surge, but also product prices and the profits of the oil companies.[16] There was a substantial absolute increase in the oil rent, and there was a change in its relative distribution, to the benefit of the companies, traders and brokers. In a historical perspective, there is a tendency for the prices of oil products in the industrial countries to follow the upward movements of spot product and spot crude oil prices, but much less their downward movements, if at all. Thus, periodic surges in spot trading and spot prices can to some extent be seen as restoring the economic margins on oil refining, trading and marketing, at least temporarily until f.o.b. oil prices catch up. Thus, surges in spot trading and spot prices can to some extent be seen as redistributive mechanisms, influencing the relative distribution of income from oil, to the benefit of intermediary traders and refiners, to the detriment of the oil exporting countries. It can reasonably be assumed that benefits in a year of crisis, like 1979, have been such that for many companies it could be of interest to possess refining capacity that shows substantial benefits only one year out of five, for example. Given the unstable political situation in the world's major oil exporting area, the Middle East, to assume that one year out of five will give a supply crisis and frantic demand with a surge in spot trading and spot prices, is not unreasonable. This could add an important speculative element to demand for oil in the world market. The spot market itself is an element of unpredictability and instability, consequently weakening the potential of the oil exporters for controlling the world oil market. This is perhaps the key reason why the new OPEC trading policy aims at reducing the significance of the spot market. However, by intending to do so, they employ a double-edged sword as they will be reducing the capacity of a safety-valve and a market regulator.

The key question is how the new state trading pattern in oil will perform in a situation of real or perceived unsatisfied demand. As a theoretical experiment, it may be assumed that by the next oil crisis, the new OPEC trading policy is fully implemented, with the oil exporting countries marketing practically all their oil themselves, and with state-to-state deals representing perhaps half the volume of oil traded internationally. The crisis is triggered off by a sudden supply cutback in one or more oil exporting countries, due to political circumstances. The remaining OPEC suppliers will then be faced with a sudden perceived unsatisfied demand, and a suddenly active spot trading. They will then face an uncomfortable dilemma; either try to dry up the spot-market by strictly enforcing the clause forbidding resale, and possibly hire a number of inspectors to supervise that refiners do not resell crude, and possibly even inspect the transactions of OPEC member countries; or make the marketing policy of oil less rigid, which means renouncing on important principles. In the first case, this perception of unsatisfied demand could be stimulated by attempts to make the spot-market dry up, which would have a most stimulating effect on spot crude prices. This could mean important windfall profits for suppliers that are not bound by the new OPEC trading policy, e.g. non-OPEC suppliers such as Malaysia, Mexico, Norway and the United Kingdom. This again would increase the temptation for at least some OPEC governments to divert some volumes of crude to the spot market. Furthermore, this would also increase the temptation for established clients of remaining OPEC suppliers to divert at least part of their crude to the spot market, and the control problems for OPEC in the current situation, with several hundred companies buying OPEC crude, are formidable, and it is fairly unlikely that an effective control could be imposed. Another element is that the surge in spot prices, even if it affected fairly minor volumes of oil, would be likely to stimulate an upward price movement and increasing spot trading in oil products, giving large windfalls profits to refiners. As far as is known, it is not part of the new OPEC trading policy to control the products sales of refiners, and the intention to forbid the resale of crude oil but not to exercise control over products sales is a serious contradiction in the new trading policy. However, to control the products sales of refiners would be even more difficult than controlling what refiners do with crude oil; in practice the task would be impossible. The OPEC countries therefore face the dilemma that if they implement the rigidities of their new trading policy in a period of supply crisis, it could have the opposite effect of what is intended, to keep oil prices from rising.[17] The alternative would be to have a more flexible marketing policy, with the ability to divert volumes of crude and products to the spot market in periods of supply crisis, in order to intervene more actively and more massively. At least in theory this could contribute to keeping oil prices from rising, or rising too high, but such a policy would require the active cooperation of the dominant oil exporting countries. Thus, the emerging dichotomy between long-term contracts under state-to-state deals and spot trading could in a period of perceived unsatisfied demand have effects quite opposite from what is intended by the new trading policy: prices could rise quickly and intermediaries could get substantial margins.

The more active intervention by OPEC governments and national oil companies in the spot market would require a recognition that the emerging market dichotomy is to some extent artificial and even harmful from OPEC's point of view. OPEC's ambition to control the world oil market is legitimate and reasonable, but the world oil market practically by definition comprises a spot or short-term market. Even if state-to-state deals should make up the bulk of international oil trade, the problem would arise, as some governments at some moments would buy too much oil and others too little. Governments may be more virtuous than companies, at least in the perspective of OPEC's trading policy, and perhaps have a propensity to balance differences between purchases and requirements through changes in inventories, but it would be both tempting and rational for governments to exchange oil and to balance surpluses and deficits among each other. Furthermore, even if the role of private oil companies and refiners has been substantially reduced, they are not likely to disappear soon from the market. Given the existence of an important private sector, it is unrealistic to assume that an intermediary market will disappear, whether for crude oil or for oil products. It would be more rational for the oil exporting countries to try to control this part of the market as well. This can be done by reserving certain amounts of crude oil, and products as the refining capacity is enlarged in the OPEC countries, for the short-term market, and to keep a certain freedom of action, so that in a period of acute crisis additional volumes can be delivered to this part of the market. This could have a negative effect upon speculation and panic, and it would allow greater flexibility, apart from the fact that a more active participation would improve information about market trends, thus improving the basis for long-term policy decisions.[18]

To conclude, the nationalization of oil export marketing constitutes a logical follow-up of the nationalization of oil extraction, but as direct oil export traders, the OPEC governments and national oil companies face some awkward dilemmas. The new trading policy is a historical innovation in the world oil market, but it contains some serious contradictions, which in a situation of perceived unsatisfied demand would become more manifest, constituting a potential threat to unity among oil exporters. These contradictions are partly linked to the asymmetrical structure of the current world oil market, with a small number of governments controlling the supply side, and with the demand side being made up by several hundred companies. In a historical perspective, this asymmetrical structure, with supply and demand being organizationally decoupled, is new, and in a more strained situation it could lead to increasing instability. For example, in 1979, under the impact of the Iranian crisis, the OPEC governments were largely outmanoeuvered by market forces and lost control over events, and were unable to pursue their aim of market stabilization. The immediate beneficiaries were, as mentioned above, companies and traders. This pattern of events could easily repeat itself. These problems could partly be overcome by a more active and massive participation by the OPEC countries in all parts of the international oil industry, pointing towards a new vertical integration with the oil exporters having a high degree of control over refining and marketing. This would give instruments for implementing a long-range price planning, and to some extent eliminate sources of spontaneous and erratic demand behaviour. Another solution would be to seek some kind of joint planning agreement with consumer governments, which would control more closely traders and refiners on their side. However, for political reasons, such an agreement for the moment seems to be fairly difficult to reach. In any case, the new trading policy of OPEC evidently only constitutes a first step, and it will be most interesting to follow the elaboration of the next steps.

NOTES

1. Fadhil J. Al-Chalabi, *OPEC and the International Oil Industry: A Changing Structure*, Oxford 1980, Oxford University Press.
2. Gudmund Hernes, *Makt og avmakt*, Oslo 1975, Universitetsforlag, p. 68 f.
3. Jens Evensen, *Oversikt over oljepolitiske spørsmal*, Oslo 1971, Ministry of Industry, p. 10 f.
4. Jean-Marie Chevalier, *Le nouvel enjeu pétrolier*, Paris 1973, Calmann-Lévy, p. 23 ff.
5. Fadhil J. Al-Chalabi, op.cit., p. 7 ff.
6. ibid., p. 3.
7. Robert Mabro, "The Role of Government in Regulating Energy Markets", paper given at OPEC Seminar, *OPEC and Future Energy Markets*, Vienna, October 3-5, 1979, p. 1-2.
8. *Le Monde*, January 17th 1980.
9. Fadhil J. Al-Chalabi and Adnan Al-Janabi, "Optimum Production and Pricing Policies" in *Journal of Energy and Development*, Spring 1979, pp. 229-258.
10. *Petroleum Intelligence Weekly*, March 17th 1980.
11. Francisco R. Parra, "OPEC Oil: Recent Developments and Problems of Supplies", *Supplement to Middle East Economic Survey*, September 24th 1979, pp. 1-7.
12. Francisco R. Parra, "Crude Sales Contracts" in *Oil and Energy Trends*, September 1980, part I, pp. 1-4.
13. ibid., p. 4.
14. Øystein Noreng, "Friends or Fellow Travelers? The Relationship of Non-OPEC Exporters with OPEC" in *Journal of Energy and Development*, Spring 1979, pp. 313-225.
15. Øystein Noreng, *The Oil Industry and Government Strategy in the North Sea*, London 1980, Croom Helm, pp. 74-75.

16. Francisco R. Parra, op. cit., p. 4.
17. ibid., op.cit., p. 4.
18. ibid., op.cit., p. 4.

PART X

Towards the International Energy Breakthrough

Chapter 22

The Growth Imperative

*Christopher Johnson**

The Venice summit meeting has recognized a fact of general importance in the solution of world energy problems, in addition to its more specific proposals. It has acknowledged that the current level of OPEC oil prices bears "no relation to market conditions". It has set up a new study of the world economy, and this could be the starting point for its discussions. If OPEC oil prices are too high, what can the rest of the world do about it?

The rest of the world has, at least until Venice, accepted rather too easily the comforting belief that OPEC is a blessing in disguise, because it has imposed on us all an oil price sufficient to call forth additional supplies to relieve the long-run shortage of energy. While this was plausible with regard to the first major OPEC increase in 1973-4, it is wholly implausible with regard to the further 140 per cent increase of 1979-80.

Even the pre-1979 price would have been sufficient to ensure the exploration and development of most of the further reserves of oil, gas, and coal—not to mention nuclear power—which are now being announced day after day.

Venezuela has, in the Orinoco reserves, about as much recoverable

*Economic Adviser, Lloyds Bank Limited.

oil as the whole of world proven oil reserves. A vast new gasfield—possibly as large as the giant Dutch Groningen field—has been found in the Norwegian North Sea. Recoverable coal reserves are 250 times current annual production, as the World Coal Study has pointed out. These are only a few examples of the plenitude of our energy resources. But it takes 5-10 years to bring new discoveries into production.

If the supply is so abundant, why does the price have to be so high? In the short run, OPEC can impose almost whatever price it wants to on the world oil market. As long as an increase in the price causes a less than proportionate decrease in sales, then OPEC can increase its revenue by selling slightly less oil at a higher price. Economic studies suggest that the sales response is only about 30-70 per cent—typically 50 per cent—of the price increase. (In economists jargon, the elasticity of demand is about -0.5, at least over a period of 3 to 5 years.)

In the longer run, the development of other sources of oil and alternative forms of energy will bring about a bigger sales response to high OPEC prices. (The elasticity of demand may be -1 or greater.) But in the meantime, how can the industrial countries prevent OPEC using its monopolistic position from continuing to raise prices, and thus increasing its members' revenue?

First, they could resolve to respond to any future OPEC price increase by arranging to cut their purchases of oil by just enough to ensure that OPEC suffered a marginal loss in its revenue. Since they agreed at the Tokyo summit a year ago to oil import ceilings, they should be able to agree to more specific OPEC oil import ceilings. If OPEC objected, it could be pointed out that the GDP loss which higher oil prices inflicted on the industrial countries would reduce their demand for oil in any case—as may be happening this year.

Second, the industrial countries could seek to produce and trade non-OPEC oil at a price below at least the highest OPEC prices, taking the Saudi price of $28 a barrel as a guidepost. OPEC now controls 60 per cent of free world oil output, but this is no reason why the rest of the world should accept OPEC price leadership, particularly that of the "hawks" in OPEC. The current high level of world oil stocks, and the slackness of demand due to the recession, makes this a good moment to try to reverse the rising tide of OPEC prices.

This might require the formation of an "anti-cartel", using the International Energy Agency. But, with some official prompting, it could be better done through the establishment of a free market in oil. The oil majors now handle far less oil trade than they used to, and some of them would be able to remedy gaps in their supplies through such a market. The Rotterdam spot market should not be allowed to have a monopoly, and there is a case for setting up a wider and more representative market for crude oil and products in the City of London.

This may seem a high-risk strategy, in view of the possibility of a confrontation with OPEC. But what is now at stake is world economic growth. It is beyond doubt that the two rises in world oil prices in the 1970s have been a major setback to growth, even if the mechanism by which this occurs is still a matter of debate among economists. Whitehall estimates at $300–400 bn. the loss to OECD real income from the 1979--80 OPEC price increases.

The most popular explanation is that rising oil prices raise the rate of inflation—typically by 3 or 4 per cent in the UK in the past year—and governments then have to deflate their economies by monetary and/or fiscal measures to get inflation down again. Another, more sophisticated, theory argues that more expensive energy permanently raises production costs, with a corresponding loss in productivity unless and until new energy-saving technology can be devised to cut energy costs back again.

Either way, OPEC has been responsible for a massive loss of world output. The less developed countries, who survived the first big oil price increase relatively unscathed, are unlikely to escape so lightly this time round. They can afford to take the loss in their stride far less than can the developed countries. The Venice summit clearly recognized their difficulties.

Since it is OPEC which has caused most of the slowdown in the industrial countries' growth rates, it is futile for OPEC to demand a real return of 3 per cent or so on the surpluses that it has invested in them. It is the rise in oil prices which has depressed the rate of return on investment, so it is perverse to expect it to be rewarded by a higher rate of return.

The advantage of a check to the world oil price rise is that, if it succeeds, it would make oil in the ground a much less tempting investment for OPEC, since its attractions depend on the assumption that oil prices will continue to rise in real terms. Once it became clear that the real oil price was not going to rise, then OPEC would have an incentive to deplete its oil faster than now. Its members might be tempted to hold production back so as to keep the price stable rather than causing it to decline. But the industrial countries could overcome this tendency by offering to guarantee a fixed minimum revenue to OPEC under long-term contracts.

The UK, as a major oil producer, is in a dilemma. We have up to now accepted OPEC price leadership—that of the higher-priced African producers in particular—for North Sea Oil. The advantages to UK tax revenue appear to overwhelm any considerations of the effect on domestic inflation.

The United States faces a different dilemma. More than any other country, it is committing itself to multi-billion dollar investments in synthetic fuels, and shale oil, many of which depend on the OPEC oil price remaining at its present level or even continuing to increase, and will even then require Federal subsidies and loan guarantees. While most conventional new deposits of oil, gas and coal will be economic at even half the present OPEC oil price, some of the fancier "synfuels" would be best kept in reserve until other, cheaper energy sources have been exhausted.

It is arguable that the benefits to the UK economy of a stable oil price would outweigh the extra tax revenue derived from any oil price increase. If the UK announced that North Sea oil would no longer follow the OPEC price upwards, but lead it downwards, the first effect might be to lower the exchange rate. This could actually increase North Sea tax revenue in sterling terms, as well as providing relief to British exporters of manufactured goods. The avoidance of further inflation from additional oil price increases would also have a beneficial effect on output and employment.

Western countries with oil, such as the United States, Canada, and the UK, should be willing to sacrifice the "economic rent" which they get in taxes from high OPEC oil prices for the sake of the wider benefits to economic growth which stable oil prices would bring, to themselves and to the rest of the world. In so far as they have already benefited from substantial windfall tax revenues they should use them for tax cuts designed to counteract the inflationary effect of high oil prices. In the UK, this would mean cutting such taxes as the national insurance surcharge and non-domestic local authority rates, which would help employers, and Value Added Tax, which would bring down the rate of inflation more directly.

If, as is widely admitted within the government, it was a mistake to raise VAT to 15 per cent, it is a mistake that can be retrieved by cutting it again as soon as the flow of North Sea tax revenue permits.

OPEC's behaviour is sometimes condoned on the grounds that OPEC is only following its members' economic interests. There is no reason why the rest of the world should not follow its own economic interests too. It would indeed be astonishing if the economic interests of the industrial countries and OPEC were to coincide, and nothing is to be gained by deluding ourselves that they happily do so.

The industrial countries are more likely to achieve their objectives by acting together, in consultation with the non-oil developing countries, than by manifesting disarray in face of a united OPEC. The Venice summit has aroused some hope that this may happen. Let the dialogue with OPEC begin. But let it temper friendliness towards vital trading partners with firmness in face of a continuing threat to world economic growth and stability.

Chapter 23

The Nuclear Route to Renewed Growth

*L.G. Brookes**

The international issues raised by nuclear energy are most usually seen as safeguarding of the use of nuclear material to prevent its diversion into military uses; and the fear that the spread of nuclear technology will itself induce or greatly facilitate the spread of nuclear weapons.

Compared with the real international significance of nuclear energy these issues—though they loom very large in any debate about nuclear energy—seem to me to be, at the most, of only secondary importance.

By far the two largest problems facing the world are:

1. The politicization of oil;
2. The threat to living standards of countries in all stages of development from an erratic and unpredictable energy market.

The two problems are of course linked. It is the great dependence of world economic, political and social order upon stable, adequate and predictable energy supplies that gives major oil producers so much political leverage.

This leverage is a very unhealthy development in world politics.

*Chief Economist, UK Atomic Energy Authority.

It forms an additional ingredient in the consideration and settlement of political problems and disputes at all levels. There may be a very good case for setting up a homeland for the Palestinians, but when a major country expresses support for it, we can never be sure whether this support would have come as easily if it did not include the bonus of gaining the approbation of the oil producing Middle East states. With so much at stake it is always possible to rationalise a stance that might not have been so readily taken up in other circumstances.

It can hardly be questioned that President Carter would have found readier and less qualified support on the Embassy hostages if there had been no oil overtones. Whether or not the Iranians have a legitimate grievance against the Americans, the action against the Embassy staff represented a breach of international law and custom of some centuries' standing. But with so much uncertainty about the level and direction of Iran's oil exports it is understandable —even though it may not be supportable—if some countries find crocodile tears to weep over Iran's role in this sorry affair.

This incident provides the link with the second of the major international problems that I identified at the beginning of this Paper. Some of the countries that are supporting the Americans— and this includes the UK—are doing so with some hesitation because they cannot really afford to lose exports to Iran in the depressed, erratic and distorted state of world trade since 1973.

I have described the effect upon world inflation and trade elsewhere (reference 1). There are three main effects from a large oil price hike:

1. The effect upon the international monetary and trading system. A system which evolved with fairly widely distributed debits and credits balancing one another out suddenly found itself confronted with all the oil consuming countries being in debt to the oil producing countries. It is then hard to approach a new balance without inducing a lower level of world trade and economic activity than would otherwise have been the case.
2. An income effect. Income that has to be spent on unavoidable purchases of oil at its new high price is diverted away from other goods and services, causing depression and unemployment in those other industries.
3. A price effect. At a simple level of understanding this simply means that producers tend to prefer less energy-intensive methods of production, other things being equal, and the

relative rise in the price of energy-intensive goods produces some shift by consumers towards less energy-intensive goods and services. This effect, may in the end, be the most important of the three.

It has been argued that the third effect is a beneficial one. It is said that it brings about a belated respect for energy, encouraging energy conservation and bringing into the market sources of energy that were uneconomic at the old price. I have dealt with these arguments in detail elsewhere (references 2 and 3). I have, for some time thought they were highly fallacious and I find much in common between my view on this matter and that of Professor Ernst Berndt of the University of British Columbia (reference 4).

But at its simplest the immediate problem that the energy consumers face is one of high prices. It does not matter much to them whether they are having to meet a high price because supply is being restricted; or whether their own demand is being choked off by a cartel-imposed high price. If high price is the problem it is hard to see how it can also be the solution. A starving Asian complaining that rice had become too expensive for him to buy for his family would hardly thank you for telling him that the price made wheat an economic proposition for him. Why should we accept the same argument for energy? New sources of energy whose cost is likely to be higher than the price that is troubling us are only relevant if there is some possibility that within a reasonable period the cost can be brought down to levels that were familiar before energy became a problem. In a short presentation there is little time to go into all the details. Some of them are set out in one of the papers that I have mentioned. I will simply repeat here Ernst Berndt's reminder, that for the most part, the economic efficiency of energy use (energy consumption per unit of economic output) can only be improved by worsening the economic efficiency with which the other inputs to production (labour, capital, materials, land) are employed. Labour can be substituted for energy at the risk of reducing output per man; capital can also be substituted for energy at the risk of reducing the average output per monetary unit spent on capital; and so on. The large increase in income per capita over the world as a whole in the last 150 to 200 years has been accompanied by the substitution of energy for all other inputs to production. We should not be surprised if physical output per capita falls when energy becomes a more highly regarded input to production. I say "physical output" because relative price changes are likely to delude us into believing that the impact of high energy prices is less than it really is. We already have one example of delusion in our midst today. In national income indices with moving bases, oil has been re-valued at a higher relative price. In other words, if we find ourselves unable to afford quite so many of the little luxuries that we could afford at one time, we are invited to console our selves with the thought that the oil we consume is worth more than it used to be. In much poorer countries the impact is at a much more fundamental level. Irrigation pumps are switched off because poor farmers cannot afford the fuel. Indians in the poor rural parts of India now spend their evenings in the dark where once they spent them in the light of oil lamps. The developed countries have experienced sharp reductions or stand-stills in economic growth.

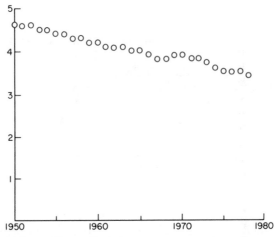

Source: National Digest of Energy Statistics.

Figure 23.1. **Energy consumption per unit of output 1950 to 1979; tons of coal equivalent per £1000 (1975) national product.**

Some of the poorer countries of the world have experienced quite sharp reduction in their already low standards of living. Will energy conservation come to the rescue? Some of its more extreme proponents argue that no other measures are necessary. I have mentioned some of the theoretical reasons why the effect may be much less than is claimed. Let us look at what has actually happened in the UK and the United States—because it is often claimed that there have been sharp improvements in energy/economic output relationships in those two countries.

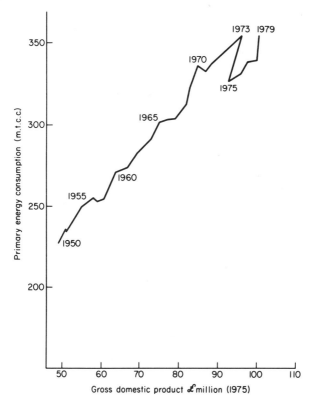

Figure 23.2. Energy consumption and gross domestic product, 1950–1979.

Figure 23.1 shows the trend of the energy/output ratio in the UK over the last thirty years. There are no signs of any change in the trend of this relationship since 1973. Figure 23.2 shows what happened to energy growth and economic growth over the same period. Both fell sharply between 1973 and 1975. Since 1975—the trough in real energy prices (as inflation overtook them) saw some resumption of economic growth and energy growth with—if anything—a steepening of the energy/GNP Plot. Figures 23.3 and 23.4 show much the same situation in the United States.

These factual graphs bear out the main conclusion in the earlier papers that I mentioned. This conclusion is that it is economic activity that bears the impact of real rises in energy prices and that the new long term equilibrium price of energy (at the new lower level of economic performance and growth) is likely to be not greatly different from what it was before the large money price hike. The

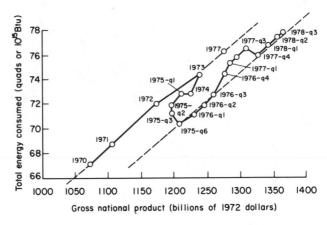

Figure 23.3. Relationship between GNP and total US energy consumption.

paper also shows that the long term equilibrium return to the oil producers is likely to be lower when prices are held up by cartel action then when they are allowed to find their own level. This is simply a corollary of the observation that the proportion of national income spent on energy tends to be remarkably constant. If income is depressed then so is the return to the energy producers.

Appendix I gives the explanation for the widely held delusion that there has been an improvement in usage since 1973.

After all this the international significance of nuclear energy should be clear. As Roger Sant, one time Washington Conservation Administrator, pointed out a little while ago: "We shall never Foreign-Policy our way out of the OPEC problem: we have to compete our way out of it". Nuclear energy represents one large new additional source of energy at modest cost. After 25 years of commercial usage it accounts for about half as much electricity in the world today as hydro power does. When all the stations under construction are completed it will equal the output from hydro power. If all the base load stations in the world reaching retiring age in the next 30 years were replaced on retirement by nuclear stations the contribution from nuclear energy could grow—over this period—to an amount equal to about two-thirds of OPEC's present entire production—even assuming no increase in electricity usage. Thus, it has the capacity to make a very large impact on the world energy market. This would reduce the political leverage of oil producing countries directly and the benefit to world economic activity that would result would go a long way to reduce economic, political and social tensions both within and between countries. It is not necessary to demonstrate that the poor countries would be able to use nuclear energy (or that ways could be found of trusting them to use it). Some of these countries are making a start towards industrialisation and higher living standards, exporting manufactured goods to the developed countries. This development has been threatened by depression in the developed countries with internal pressures to restrict imports. The higher levels of world economic activity and easier energy prices that would result from a new large injection into world energy supply would benefit all the countries in the world. If this and other opportunities—such as the establishment

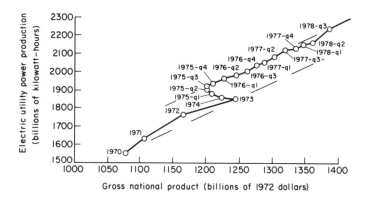

Source: Bechtel.

Figure 23.4. Relationship of GNP to electricity product in United States.

of a world trade in coal—are pursued it would probably do more than any other single course of action to relieve international tension and bring about a happier international order. The OPEC countries themselves would also benefit. They have their own problems. They are keen to diversify their economies against the day when their oil supplies start to dwindle. More predictable energy prices and a more prosperous world economy give them both a stable market for their present principal product and a more prosperous world in which to diversify. The alternative may be a saw-tooth pattern of world performance, limping from one recession to another towards ever more dependence upon fuel with political strings attached to it.

APPENDIX I

Why the Energy Coefficient is an Unreliable Measure of Changes in National Energy Efficiency

The energy/output ratio has been falling at the rate of one per cent per annum for at least the last three decades. (The reasons for this need not concern us here.)

Over the same period the G.N.P. growth rate was about 2.75 per cent per annum up to 1973 and low and erratic after that.

Thus, up to 1973, the energy coefficient was:

$$\frac{2.75 - 1.0}{2.75} = 1 - \frac{1.0}{2.75} = 0.64 \qquad *$$

Generalising, if g per cent is the G.N.P. growth rate, the energy coefficient over the whole period becomes:

$$1 - \frac{1.0}{g} \qquad *$$

Which clearly gets smaller the smaller g becomes, falling to zero when g falls to 1.0 per cent—the long term annual improvement in the energy/output ratio—and becoming negative when g falls below 1.0.

The low and erratic value of g since 1973 explains why the energy coefficient has been low since that date although there has been no change in the trend of the energy/output ratio.

The energy/output ratio always measures national primary energy efficiency in terms of the number of units of energy consumed per unit of G.N.P. no matter what is happening to the G.N.P. growth rate.

REFERENCES

1. "The Nuclear Power Implications of O.P.E.C. Prices". By L.G. Brookes. *Energy Policy*, June 1975.
2. "The Energy Price Fallacy and the Role of Nuclear Power". By L.G. Brookes. *Energy Policy*, June 1978.

*Strictly speaking growth rates cannot be added and subtracted in this way unless they are true exponential rates. The error introduced with small percentages like this is, however, negligible.

3. "Energy, Inflation and Economic Prospects". By L.G. Brookes. Paper to "Design '79", a conference organised by the Institution of Chemical Engineers at the University of Aston in Birmingham in September 1979.
4. "Energy, Capital and Productivity", by Ernst Berndt and Cathy White, presented at the annual conference of the I.A.E.E., Washington June 1979.

Chapter 24

The Realities of the Energy Market, an Agenda for the 1980s: Decisions and Research**

*M.A. Adelman**

Oscar Wilde defined a cynic as one who knew the price of everything and the value of nothing. Such a person is not fit to make decisions on public policy, but he is a useful servant. The economist is a professional cynic. He should know what things cost, what society pays for a sulfur standard of 1 rather than 4 percent, and why going from 1 to 0.5 percent multiplies that cost. It is not so clear that the economist, even if he knew how many lives would be saved or improved by less sulfur pollution, has special competence in deciding whether their value exceeds the cost.

In working such problems, we must be able to understand information and questions addressed to us by engineers, lawyers, business managers—and by what my teacher Schumpeter called "that refractory wild beast, the politician." The terrain we work with them is the study of energy markets.

Any mention of markets sets off, I fear, a set of conditioned reflexes. The problem is not new. Edmund Burke once pronounced:

*Massachusetts Institute of Technology, Cambridge, Mass., U.S.A.

**In so brief a review, it is inexpedient to include references. But I obviously am indebted to more writers than I could cite, as well as to the comments of James Paddock, Robert Pindyck, and Martin Zimmerman.

"The laws of commerce are the laws of Nature and therefore the laws of God." Karl Marx retorted in his irascible way: "No wonder that in accordance with the laws of God and Nature, [Burke] always sold himself in the best market." Marx's comment was unpleasant, but Burke was doing mischief presenting a cardboard stereotype of The Market, a marvelous machine to dispose all for the best. The opposing stereotype is The Market as not ordained of God but contrived by the Devil. Let us put "The Market" aside. Markets are many. Some are purely competitive and work fairly well—or not well, for lack of information. There may be increasing returns, hence natural monopoly. This was supposed for years to be true of petroleum—which turned out to be a strong case of diminishing returns. Or monopoly may be deliberately achieved by public or private bodies. There may be enormous externalities or side effects, chiefly damage to the environment. Our business is to ascertain how particular markets work, and with what consequences.

Markets are worth attention because they are usually out of equilibrium. Hence the central concept for studying them is equilibrium—a good theory is not a description. We need to see which way a market is heading, and what factors, conventionally grouped under Demand and Supply, will disturb them Further. That sets our agenda for the 1980s.

ENERGY DEMAND

The price explosion of 1973-74 ushered in a panic which economists did little to cure. Public and official opinion is still seized of the idea that to exist and grow we must have "enough energy for our needs", a fixed quantum. Hence the irrelevant polemics of growth versus anti-growth. Or at a slightly higher level of sophistication, "the demand for energy is inelastic". This statement is not useful, not even meaningful.

Energy demand in any given country at any given time is ruled by three factors: the gross national product; energy prices relative to all other prices; and whatever is not subsumed under the first two: call it "structure". For example, urban sprawl, with heavy consumption of energy for home heating and transportation, goes far back into the past. Over the years, I have risked the derision of my students in telling them that 350 years ago people migrated from Old England to New England—of all places—in order to stay warm. But the documents prove it: "All England, nay all Europe, hath not such great fires as we have in New England. ... Here is good living for them that love good fires." That was the obvious effect of relative prices, but it set in place a certain residential pattern which may long survive the original reason, and is resistant to change.

Prices work through substitution, and there are three substitutes for energy: (1) labor, (2) capital, and (3) everything else. Lest you think class (3) is slightly flippant or oversimple, consider only automobile usage. Motorists do not "need" more fuel-efficient vehicles. They would rather spend some of their limited incomes on more of other goods and services, and less on motor fuel.

Relative prices act like a glacial drift—imperceptible in the short run, irresistible in the long run. Since changes take time, the system is always out of equilibrium. In 1973–78, the average increase in real price to household and industrial consumers in the OECD was roughly 40 percent, mostly coming right after the 1973 production cut, the so-called "embargo". During 1973–78, the reduction in OECD energy use, per unit of gross national product, was about 9 percent. (The less developed countries have not reacted as strongly: a temporary respite from the need to adjust, made possible by heavy borrowing.) Assume that the biggest savings are made earliest so that the rate of response keeps slowing down. Assume also that the average reaction time is somewhat less than the average lifetime of the stock of consumer and household capital. At a rough half-life of seven years, perhaps 40 percent of the eventual reaction had been accomplished in five years, with a greater reaction still to come. Such a trend would by 1990 lower the energy used per unit of GNP by 20 or more percent in the OECD countries.

The 1973–78 decline in energy use seems consistent with econometric analyses of consumption across countries, sector by sector, showing elasticities in the range between 0.5 and 1.0. But these results are highly tentative. Aside from many econometric traps and snares, one must also test its basic assumption, that the variation among countries at any given time is a good enough proxy for long run demand elasticitiy.

There is a complementary type of inquiry. A study, completed before the latest price rise, concluded that given the price-cost relations of 1979, 30 percent of energy used in Western Europe could be saved, by appropriate investment, at a profit to the user. This 30 percent saving is in addition to the 9 percent already saved, suggesting that the ultimate reaction to the price rise will be greater than just suggested: again, given time to reach equilibrium. A similar study for the United States shows a lesser adjustment: as of 1978, about 20 percent of energy use could be saved, at a profit on the needed investment.

However, capital and energy are both substitutes and complements, and pre-1973 relations are plausibly explained stressing either one. How do we devise ways to give substitution and complementarity their respective dues? Perhaps the econometric puzzles of the pre-1973 experience will be solved because we have now accumulated five years of later history, which need intensive analysis. Atop that, there is the 1979-80 price rise, to which the reaction has hardly begun.

This second round raises some basic questions. To say that demand elasticity "is" one or another figure, is to talk very loosely. As noted earlier, for any set of prices, one gradually exhausts the easy and cheap ways to save. But the higher the price of energy the larger the investment which is worth making in order to save energy costs. The second effect should overbear the first, and elasticity should be greater with higher prices. But how much greater, in what country, and in what sector of energy use, with what substitution patterns?

Two final reflections on demand. First, energy savings may mean using factors much less productively. Is it only coincidental that productivity—output per man hour—has done so poorly in practically every OECD country since 1973, in contrast to the variety of favorable trends in earlier years? Some would explain this by "inefficient" substitution of labor for energy, and would expect matters to improve by the slower substitution of capital for inefficient labor.

Second, the OPEC special committee on long run price policy (chaired by Sheik Yamani) expects to keep increasing oil prices to the ultimate ceiling, which they consider the supply price of synthetic fuels. That is logical: since they will no longer permit oil to compete with oil, the first point of competition is with substitutes. The demand for oil would turn very price-elastic in the neighborhood of the synthetic oil cost at which very large volumes would be available. Sensible monopolists will not exceed such a price, since they would lose so much sales volume as to lose net revenues.

The Committee's approach is therefore rational. But it is incomplete. What if demand gets so elastic, even before synthetic parity is reached, that losses would be suffered by increasing the price? The real price as consumed of a barrel of products in Western Europe this year will be rather more than twice what it was in 1973. Hence the reaction to additional price increases should be stronger. Furthermore, in 1973, the crude oil price was about one fifth of the product price, and OPEC oil was nearly two-thirds of world non-Communist consumption. Hence elasticity of demand for OPEC crude oil in 1973 was only about (0.2 / .64) 30 percent of the elasticity of demand for products. In 1980, the crude oil price is about 60 percent of the product price, and OPEC oil is just under 60 percent of non-Communist consumption, so demand for OPEC crude is now about as elastic as for products generally. Therefore, an OPEC crude price elasticity which was so small in 1973 that it was lost in the noise of the system is today enough to reckon with.

How far is the price today from its ultimate high? The OPEC nations would be acting rationally in raising prices somewhat higher than the level where long-run elasticity exceeds unity, and long-run revenues are maximized. This is because the temporary higher revenues come soon, while the long run lower revenues come later. I happen to think there is still room for further increase. But my opinion aside, those time lags mentioned earlier, give the energy markets a strong tendency to overshoot, to everybody's loss. Great turbulence is inevitable. More knowledge of demand, acquired and diffused in good time, might lessen it.

GOVERNMENT AND ENERGY DEMAND

Governments are not a force extraneous to markets. They may be sellers, or more rarely buyers. But they may also intervene in accordance with the perceived interests of buyers or sellers. Economists have long agreed on a policy maxim: since a price is both a means of guiding resource use and of income distribution, let the solution, like the problem, be two-phase. Let the price be such as to maximize efficiency, which increases total income; then distribute income according to what is perceived as justice. The advice is rarely accepted, so we must analyze consequences. For example, had American consumers been confronted with the fact of higher oil prices six years ago, they would have straggled toward more fuel-efficient automobiles. Having been hit belatedly, they are now stampeding that way, and Detroit is thrown into convulsions trying to adapt.

A very different demand policy is mentioned by the head of the New York Federal Reserve Bank: "instead of paying increasing taxes to OPEC . . . we can pay taxes to ourselves and recycle the proceeds domestically." How soon the importing countries will consider pre-emptive taxation may be the most important demand forecast we could make for the 1980s.

THE WORLD OF SUPPLY

The demand reaction to price changes is relatively discernible and predictable because it sums up the decisions of millions of firms and households, each comparing an energy price with the prices of alternatives.

The world of supply is much more difficult to understand, even if we were to assume markets are all working competitively. For example, although commercial nuclear reactors have been operating for years, I wonder if anyone is yet in a position to estimate the steady-state cost of nuclear power under current technology. I remember two students telling me, early in the 1970s, that they had spent a fascinating summer working with problems in a large commercial power reactor. It struck me at the time that if these talented young people found so much that still needed study, operating that reactor was costing much more than its owners had bargained for. It never occurred to me to generalize that incident. But do we know enough today to make a rational choice between nuclear and coal-fired power in most locations throughout the world? Such an enquiry would need to separate the costs of uncertainty (about government policies and procedures) from the uncertainty of costs. In my country, I suspect, no new nuclear power plants will be started, for the time being. In France, just the other way. In neither country, can I discern a weighing of perceived costs and benefits.

But the uncertainty over nuclear power operating costs is less than that in energy minerals: coal, oil, gas, and uranium. Now, the significance of any supply curve depends on expected demand. It seems like yesterday that uranium was expected to become superfluous early in the next century as fuel reprocessing and the breeder reactor squeezed much more of the potential energy out of each pound. The wheel has been coming full circle: breeding and reprocessing receding; fear of uranium scarcity; a wild run-up in prices; large if undefined new resources identified; prices starting to collapse, and I care not to predict the next scene. But the moral is there: the demand for a mineral depends on current and expected technology in consumption.

This brings us to the central core of mineral economics, the proposition that under stable conditions, and abstracting from cost, today's price would equal the future price discounted at the relevant discount rate. It was an immense contribution to show that future scarcity would force up prices today, and that mankind is not in danger of driving blindly off the cliff of apparent plenty into the void of materials running out, because it receives signals a long way ahead.

But as noted, a good theory may be a bad description. With a given mineral stock, if the real discount rate were, say, 3 percent, minerals prices would have to increase by a factor of nearly 20 in a century. But real minerals prices (monopoly aside) have if anything tended to decrease, while the prices of the "renewable" resource timber has been rising. This makes us reverse the question—not whether rising prices cast their shadow before, but why prices should rise.

Minerals production, as remarked earlier, is a strong case of diminishing returns. The better prospects are explored and developed first. When a prospect becomes a mineral body, its ever-more intensive exploitation becomes ever more costly. Hence marginal development costs rise over time, and so must price, all else being equal.

The anticipation of higher costs, hence higher prices, promises higher future royalties on any existing deposit. Those higher future royalties, discounted to the present, are *ceteris paribus* a part of the cost of using up the stock today. But the other things need not— almost surely will not—be equal. The expected higher price slows consumption, hence slows the rate of increase of marginal costs. Higher prices also generate the search for new deposits, and for more efficient extraction, which may slow, or even reverse, the otherwise inevitable rise in development costs. At the margin, the cost of developing a unit of capacity in a known deposit equals the cost of finding-cum-developing a new deposit; hence development costs are a proxy for exploration costs, and trends therein. Monopoly aside, rising development costs are not merely crucial, they are the whole process of price determination. All the more regrettable, therefore is our scanty knowledge even of current development costs in oil, gas, coal, and uranium; and the sketchy data we have on "reserves", where concepts and data vary from country to country, and rarely meet even minimum standards of meaning, let alone accuracy. Yet the foundations of any policy of substituting, among coal, oil, gas and uranium, should be comparisons of long-run marginal development cost. Great hopes and fears, and grand complex models, ignoring costs, remind one of a Cambridge luminary, Bertrand Russell, defining pure mathematics: the field of study where you never know what you are talking about, nor whether any statement is true or false.

All research is perpetual oscillation between methods and data with an improvement on one side making the other bind. In my judgment, we need data more than better methods. In the physical sciences, large costly projects are designed and built to generate observations. Analysis is often quite brief. In economics, and no less energy economics, we scavenge for bits of data and put them through complex number-crunching, with naturally dubious results.

In summary, the basic difference between demand and supply, in my view, is a difference of degree so great as to be a difference in kind. For demand, relations drawn from past experience will to a great extent hold in the future; econometrics, which is merely an orderly summation of the past, is a predictive tool. It has not served us on the supply side because it does not capture the peculiarities, as sketched here, of a depleting resource, and the reactions thereto. Still less does it capture the peculiarities of government management of energy supplies, of which more later.

At this point, it is necessary to say more about the discount rate governing energy supply. All decisions are made in the present to commit the future, and discounting makes the present comparable with the future, toward which it moves on a disturbed path.

Now consider the widely held belief that private owners of energy minerals tend to produce them too quickly, at the expense of future production, because those owners' risk aversion makes their effective discount rate higher than it would be in a frictionless world, where they could insure or hedge themselves against risk.

It seems to be generally assumed that higher discount rates, particularly higher risk allowances, speed up extraction. In fact, they may slow it down, and very high risk infallibly slows it down. The mistake lies in ignoring the role of investment. Energy production is capital intensive, not only involving impressive physical structures, like offshore oil drilling platforms, but often long lead times, requiring intangible but costly investment. Higher discount rates invite faster disposal of the capital asset mineral-in-the-ground; but they simultaneously increase the opportunity cost of investment in finding and extracting those minerals. Which effect overbears which depends on the required relative proportions of the two types of capital; what slows things down in a high cost deposit speeds them up in a low-cost deposit.

We can look at it a bit more directly. If "risk" is a perceived high probability of an outside shock, like expropriation, it is a damper on investment, hence it slows down the rate of mineral extraction. Or "risk" may indicate great uncertainty around an expected positive net return. Were there no variance, there would be investment. But things may turn out worse than expected, a reason for holding back, not investing. And since more may be learned in time, it may be worth waiting to learn. Hence the greater the variance around the expected mean value of the investment, the more reason to wait. The effect is, again, to delay investment, hence delay extraction. Both kinds of risk are depressing investment today.

The discount rate is part of the basic decision of every mineral operator since the world began: what is the optimum depletion rate? There is a trade-off: more intensive exploitation raises marginal cost, but brings receipts in sooner. The optimal rate puts net present value to a maximum. If faster depletion would increase the present value of the deposit, the depletion rate is too low. If slower depletion would increase present value, the current rate is too high. The problem, like most economic problems, is more or less, not yes or no. Therefore the popular slogan that "oil (or gas or coal or uranium) in the ground is worth more than money in the bank" is not wrong; it does not rise to the scientific dignity of error; it is nonsense.

So much for our problems with the peculiarities of a depleting resource even under competitive conditions. We turn now to non-competitive energy markets, which seem mostly to be a special case of government management.

THE INFLUENCE OF GOVERNMENTS ON ENERGY SUPPLY

The influence has been so great as to reverse normal relation-ships, and this is an additional reason why econometrics are no help on the supply side. In many areas, for divers reasons, the higher the price, the less the investment: a backward-bending supply curve.

Start with governments in their traditional role as landlords or concession-grantors. In most of the world, a good discovery means a dissatisfied landlord who will probably use his sovereign power to tear up the contract. The higher the price, the greater the inducement to revise unilaterally. The World Bank is today grappling with the task of trying to restore some of the system of contracts.

In some consuming countries higher energy prices mean "obscene" profits to some producers, hence price ceilings and taxes, and a reluctance to lease public lands for coal or oil development. In the United States, the greatest untapped oil resource is what has been left behind in old fields. Costly experiment to devise and test new recovery methods was not worth doing under price ceilings which were in real terms lower than in 1973. Investment in enhanced recovery has if anything declined. Possibly price decontrol subject to the new misnamed "windfall profits tax" has provided an escape hatch. In Canada, very large deposits of heavy oils, profitable at the price of three years ago, are in limbo because the Federal and Provin-cial Governments each assert a claim to part of the profits, which claims add to over 100 percent. Were the prices of oil and coal lower, public opinion would not so urgently require these policies.

Minerals including energy are often considered a treasure too good to share with the un-deserving foreigner. Again, the higher prices go, the deeper bites the Club of Rome delusion, that we are running out of everything, and had better clutch the stuff to our bosom, rather than find, develop, and produce it.

Among the members of the great cartel of oil producing nations, the greater their revenues, the more effective their cohesion, since the pressure is less on any to seek higher revenues by shading prices. The higher prices go, the higher they go. Again, for quite a different reason, there is a backward bending function.

Some would suggest even more unusual patterns. Adam Smith was very skeptical of those who professed to be in business for the public good. But "we have changed all that". The largest oil producer nations are said to be flouting their own economic interests. For the public good, they produce at higher rates than optimal for them. They do not aim at present values, but only to satisfy "needs". That was the dominant view in 1973, when, for example Saudi Arabia was realizing about $14 million daily: they would prefer lower revenues. But the new Saudi Five Year Plan requires $164 million per day, in real terms roughtly six times as much as 1973. This covers only public domestic investment, not public consumption, nor public foriegn investment, nor the whole private sector. So total "needs" have expanded by a factor of six to ten, at least, in 7 years. Perhaps, like the rest of the human race, Saudi revenues determine "needs". The simpler hypothesis ought to be preferred to the more complex: the special story had better be a good one. Perhaps the hypothesis that Saudis *et al* produce more than suits their economic interests requires some evidence.

The oil producer nations' cartel is an all time success, but there is or has been, a uranium group whose doings are still obscure, not to mention coal-producing States of the United States, who dream dreams and see visions.

Those controlling the State may wish to build up absorptive capacity, meaning mostly trained people; or to ward off inflation; or limit the number of foreigners to be allowed into the country, etc. Such considerations push toward more foreign less domestic investment. These factors are mentioned all the time, but it is not so easy to cite careful analytic works of how they impinge on energy production.

Private corporations try to maximize their owners' wealth; producing governments aim to maximize their constituents' wealth. But who are their constituents? Perhaps it does not much matter: perhaps obtaining wealth is a task independent of its distribution. But one group or faction may have a different idea of what maximizes wealth. To sharpen the inquiry we might ask: would a government deplete an energy mineral faster or slower than a private concern?

Governments are said to have lower discount rates, which could as seen earlier speed or slow depletion. Governments are also said to have longer time horizons, which can act like lower discount rates. But this is usually not true. A tyrant, a dynasty, a party in a one-party state, a party watching the next election, a faction, a junta—these people do not need my advice to have a relatively short time horizon. Again, it is a function of time and place.

Take one special but important case. Where reserves are very large, the alternative to producing a unit today is to produce it in 30 or 50 years, at negligible present value. This makes no sense in a private or government producer acting independently of its rivals. It makes excellent sense as a means of restraining output to keep up the price.

"Conservation of natural resources", as practiced for many years by the oil producing States of the United States, meant regulation of output to raise and maintain prices. There was discord aplenty among Texas, Louisiana, and the others (as among Saudis, Kuwaitis, *et al*) but harmony prevailed on the whole.

Economists who analyzed the meaning of "conservation" became highly unpopular. The then Governor of Texas, John Connally, honored me with a personal denunciation. A great Cambridge economist, Alfred Marshall, once warned that economists were in trouble when everyone approved of them. I hope we are in no immediate danger.

Other kinds of government policy may also be important. The limited monopoly of the coal carrying railroads in the United States is strengthened and enforced by the government, to the point where the good people of San Antonio, Texas, are seeking coal from Australia because the railroad bringing in coal from the Rocky Mountains has overshot in using its price raising power. It all follows from a set of rules meant to insure "fair" prices. Railroads in Australia are publicly owned, and set rates about the same way.

The impact of other government actions is more difficult to discern, and merits close study. For example, the last two years have seen the near-complete ouster of the multi-national companies as the direct sellers of crude oil, but they remain as refiner-marketers. Crude oil now follows many new and indirect mid-channels. Is it, I wonder, easier or harder to control a motley crew of middlemen, all of whom must do more buying and selling than did the multinationals? Are we on the eve of developing a set of spot markets and and term markets in crude oil similar to what already exists for many commodities, and to an important though incomplete degree in coal and uranium?

This brings me to my one policy suggestion. In the long run, "access" to crude oil, "availability" of supply, are non-problems; the market will equate supply and demand, at a price which will please some and grieve others. In the short run, the danger is real: not of a cut-off, but of a cutback. Oil may again be used for political ends. More important, the producing nations are unstable. Some are colliding painfully with the twentieth century, and with few exceptions, the "Outs" can only replace the "Ins" by conspiracy and violence. Most important, these governments are fixing production volumes, rather than fixing price and then supplying whatever is demanded at that price. Their output must be higher or lower than what is demanded, even in total. The discrepancies get much worse if we look at particular types and locations of crude oil. Hence there must always be a good chance of glut or shortage. But while the producing governments can endure (or enjoy) a shortage, they abhor a glut. Hence the odds are far from even, and the very fear of shortage makes demand surge, provokes shortage, then panic. The result is our old friend consumer surplus, Dr. Jekyll suddenly turning into Mr. Hyde. There is a remarkable difference between the price of oil, however high, and the damage inflicted by not having any. Hence the rational but very disruptive urge to hoard.

The only solution is strategic storage, which exists nowhere today. (The E.E.C. ninety days is mostly illusion because it is mostly pipeline fill and tank bottoms.) A United States law providing such a reserve is five years old; but President Carter has declined to enforce it, without the royal assent of the King of Saudi Arabia. Congress has overruled him in principle, but only to the extent of 36 million barrels per year, which will attain the target in 2000 A.D. If consuming-country governments ever summon up their courage and good sense, they will build strategic storage, and for its use they will borrow from Victorian finance theory. Walter Bagehot laid down the rule: in time of panic, lend freely, but charge the earth. Governments should sell crude oil freely, at the cost of storage plus the highest price being paid anywhere in the world. Buyers need no longer fear dearth—wheels stopping, people thrown out of work, freezing in the dark, etc. Since the price anticipates near term expectations, few will be tempted to buy in order to speculate. A temporary excise tax would restrain demand and mop up windfall gains, and should be terminated quickly, not phased out.

There is no sign that our governments understand this: perhaps some of us will help persuade them.

Annexes

DETAILS OF FULL PROCEEDINGS
(AND ACKNOWLEDGMENTS)

The 1980 Annual Conference of the IAEE

The decision to hold the second annual Conference of the International Association of Energy Economists in Churchill College, Cambridge, UK, in June 1980 had its origin in talks between James Plummer, President of the Association 1979/1980 Jane Carter, then Chairman of the UK Chapter, and Professor Robert Deam of Queen Mary College, London. Thanks to their foresight, enthusiasm and guidance, it was possible to announce firm details of the Cambridge Conference one year in advance at the Washington Conference in June 1979.

The planning and organisation of the Conference was shared by many members. The Programme Co-Chairmen were Jane Carter (of the UK Department of Energy) and Professor Ernest Berndt, then of the University of British Columbia. In the United States, James Plummer co-ordinated the US speakers, while William Hughes took charge of the complexities of handling the registrations in the United States. The Conference Co-Chairmen were Paul Tempest and Richard Eden (Director of the Cambridge Energy Research Group). Under their general direction, the Conference Planning Committee met regularly for a year at the Cavendish Laboratory, Cambridge, and the smooth running of the Conference owed much to Anne Swinney and Cynthia Wilcockson of the Cavendish and to Anne Ffowcs

Williams and Sandy Gordon of Conference Associates (Cambridge).

In Cambridge, we were particularly indebted to the Master of Churchill College, Sir William Hawthorne, for extending to us the very warm hospitality of his College and the considerable benefit of his support and enthusiasm. A major contribution to the enjoyment of the Conference were our two distinguished after-dinner speakers, Sir Jack Rampton, Permanent Under-Secretary at the UK Department of Energy and Michael Posner, Chairman of the Social Science Research Council.

The Chairmen of the ten plenary sessions and four parallel sessions gave generously of their time in the co-ordination and introduction of the speakers and in summarising their remarks in the final round-table discussion. Above all, the continuous open debate both within and outside the formal sessions made the Conference a memorable, useful, enjoyable occasion.

Since the Conference, the IAEE has continued to expand to the 2,000 member mark under its new Chairman, Professor Adelman of the Massachusetts Institute of Technology. The UK Chapter has reconstituted itself as the British Institute of Energy Economics. Other national chapters are now being established in Japan, Germany, France, Norway, Canada, Venezuela and India, with a system of national representatives in other countries. Conferences have been arranged in Washington (October 1980), Toronto (June 1981), Houston (November 1981) and Cambridge, UK (June 1982).

International Association of Energy Economists
Opening by

Professor Sir William Hawthorne

Mr. President, Ladies and Gentlemen, it gives me great pleasure to welcome this distinguished Association to Churchill College. As you know Sir Winston Churchill founded this College, gave it his name, and as Chairman of its Trustees attended every meeting until he became ill. He chose to have a College as his national memorial and not a palace like Blenheim. I hope, however, that although the accommodation is not palatial, you will still find it comfortable. I see that there is an excellent description of the College in the programme. In case you do not know where you are you will note that on the back page it says we are in Cambridge, U.K. I should perhaps point out that the picture on the front is not of our chapel. The chapel at this College has a somewhat different style.

In welcoming you to the College I also welcome you to the University. The University is of great antiquity but moderate size, it consists of 94 departments and innumerable laboratories, libraries, workshops, farms, gardens, printing presses, telescopes and computers. The 8,000 undergraduates are about equally divided between the Arts and Sciences. In this University for some reason we regard Economics as an Art—behind its mediaeval facade—the University's I mean several thousand graduate students, research workers and staff advance knowledge in their subjects ranging from molecular biology to radioastronomy, from plant breeding to computer science and, of course, from economics to energy.

The fact that energy and economics is not only the purpose of our meeting today in Cambridge—with its Bridge of Sighs—but also that of a meeting in another ancient city with a similar bridge must surely indicate the importance of the topic. It is not necessarily a new one, for engineers have long been concerned with the economics of energy use. One of the most famous, James Watt, sold his steam engines in 1775, by taking a 'premium' or royalty fixed at one third of the fuel savings obtained by his invention of the separate condenser, an invention which trebled the efficiency previously obtained by the obsolescent Newcomen engines. Over the years there has generally been a steady improvement in the efficiency of energy use in response to economic pressures. But such pressures have been variable and sometimes ineffective. Worse still, there are some countries which for various reasons choose to prevent their citizens from receiving the same economic signals as the rest of the world. Today the problem of supplying the world's growing demand for energy has become serious to some extent because of the large quantities involved. We are using twelve times as much energy as we did in 1900 and twice as much as we did in 1960. Also, if we want to use more, it is not likely that we will get the more from petroleum or natural gas. And whether we find the more from coal, nuclear or other sources or can reduce demand by conservation the lead times are now much greater than they were in James Watt's day.

This conference has a chance to study and debate all such features of the energy problem, to tease out the vagaries of pricing policy, or in taxation, for instance why only tax the use of oil in the transport sector when there is as yet no substitute for oil—to weigh the energy models, and to reassess the energy scene—which since the Association last met—has been changed by a rise in price of oil greater than that of 1973.

You have arranged a very extensive programme and I am sure you wish to get it started as soon as possible. So—once more may I welcome you to Churchill College and wish you well in the challenging task you have set for your Association.

Programme

SESSION 1 (Chaired by Sir Jack Rampton)

Key International Energy Problems of the 1980s

1   Risks in the International Oil Trade
     Sir David Steel
2   Energy at the Crossroads: Abundance or Shortage
     Chauncey Starr
3   Key International Energy Problems of the 1980s
     Melvin A. Conant

SESSION 2 (Chaired by Sam H. Schurr)

Comparisons of World Energy Forecasts

1   World Energy Prospects
     John Foster
2   A Comparison of World Energy Forecasts
     A. Harper and J.V. Mitchell
3   The IIASA Global Energy Study
     Paul S. Basile

Francisco Parra (Director, International Energy Development Corporation), *James Hanson* (Chief Economist, Exxon) and *James Reddington* (Economist IEA) also contributed to the second session.

SESSION 3 (Chaired by James Plummer)

International Comparison of Governmental Responses

1 A Conceptual Framework for Examining Governmental Responses to Energy Vulnerability
  James L. Plummer
2 Governmental Responses to Energy Problems in the 1980s
  John C. Sawhill
3 General Observations regarding Governmental Response to Energy Problems and Issues
  Ali A. Attiga

A. Parra (Petroleos de Venezuela (USA) Corporation) and *Ian Smart* (Ian Smart Ltd) also contributed to the third session.

SESSION 4 (Chaired by Richard Eden)

Energy in Developing Countries

1 The Mexican Oil and Gas Sector—Selected Statistics
  Adrian Lajous-Vargas
2 Oil and Energy Demand in Developing Countries
  James Navarro, Charles Wolf, Jr and Daniel Relles
3 Financing Energy Developments in the Third World
  R.K. Pachauri

SESSION 5 (Chaired by David Sternlight)

International Energy Production and Trade

1 International Oil Production
  Peter R. Odell
2 East-West Factors in International Energy Production and Trade
  M.C. Kaser

3   International Developments and Trade in Uranium
     Joen Greenwood
4   Long-range Pricing of Crude Oil
     R.J. Deam

SESSION 6 (Chaired by Jane Carter)

Estimation of Demand and Conservation

1   The Impact of Crude Oil Price Rises on Oil Product
    Consumption
     A.F. Beijdorff
2   Estimation of Demand and Conservation: the Developing
    Countries
     Joy Dunkerley
3   Aggregate Elasticity of Energy Demand
     James L. Sweeney
4   Energy-Output Coefficients: Complex Realities Behind
    Simple Ratios
     G.C. Watkins and E.R. Berndt

SESSION 7 (Chaired by Paul Tempest)

Institutions and Policies for Managing Energy Shortages

1   Institutions for Managing Energy Shortages
     Fritz Lücke
2   Managing Oil Contingencies: the US Experience and
    Choices for the Future
     Michael L. Telson
3   State Oil Trading and the Perspective of Shortage
     Øystein Noreng

Edward Krapels (Consultant, USA) also contributed to the
seventh session.

SESSION 8 (Chaired by Jeffrey Lewins)

International Nuclear Issues

1 The Nuclear Nonproliferation Act as a Constraint on
  Foreign Policy
  Charles K. Ebinger and William G. Young
2 France's Nuclear Power Programme
  A. Ferrari
3 International Nuclear Issues
  L.G. Brookes

Ian Smart (Ian Smart Ltd) and *Irwin Bupp* (Harvard Business
School) also contributed to the eighth session.

SESSION 9 (Chaired by Christopher Johnson)

Energy Prices and Productivity

1 Energy Prices and Productivity Growth
  Dale W. Jorgenson
2 The Growth Imperative
  Christopher Johnson

SESSION 10 (Chaired by Richard Eden)

Political Factors in Energy

1 Trends in the Next Ten Years
  Norman Lamont

Robert Mabro led the round-table discussion, chaired by
James Plummer, which followed the tenth session.

CLOSING ADDRESS

An Agenda for the 80s—Decisions and Research
M.A. Adelman

PAPERS FROM PARALLEL SESSIONS

PARALLEL SESSION A   (Chaired by Alan Manne)

Models and International Comparisons

1 A Three-Region Model of Energy, International Trade and
   Economic Growth
   Alan S. Manne
2 The Value of Time in Powerplant Licencing Procedures
   Gunter Schramm
3 Industrialisation, Energy Requirements and the Quality
   of Life: an International Comparison
   Ben-Chieh Liu and Claude F. Anderson
4 Social Opportunity Cost of Electricity
   V.S.S. Suresh Babu
5 Forecasting the Business Environment in an OPEC World
   Anthony J. Finizza
6 Endogenous OPEC Behaviour and Consumer National
   Policies
   Richard F. Kosobud

PARALLEL SESSION B   (Chaired by Keith Williams)

Energy in Developing Countries

1 The Role of Nuclear Power in the Third World
   Marcus Fritz
2 Nuclear Power for Developing Countries
   Eli B. Roth

PARALLEL SESSION C   (Chaired by Terry Barker)

Energy Pricing and Conservation

1 Governmental Subsidization of Domestic Energy Supplies
   J. Rodney Lemon
2 Pricing Policies for Cogeneration and Other Small Power
   Production
   Patricia Davis and Jackalyne Pfannenstiel

ANNEX IV

IAEE and BIEE Committees

INTERNATIONAL ASSOCIATION OF ENERGY ECONOMISTS

President: M.A. Adelman
President-elect: William R. Hughes
Immediate Past President: James L. Plummer
Executive Vice President and Secretary: Joen Greenwood
Treasurer: Brian Sullivan
Vice President for Publications: Joy Dunkerley
Vice President and Newsletter Editor: Linda Ludwig
Vice President for International Affairs: Jane Carter
Vice President for Chapter Liaison: Larry Blair
IAEE Council: Marino Gurfinkel, Linda S. Rathbun, Michael Telson, Paul Tempest, Arlon Tussing

The International Association of Energy Economists was founded in 1978 in the United States and now has over 2,000 members. National chapters are being formed in almost all major countries. The United Kingdom chapter was established in 1978, changing its name in late-1980 to The British Institute of Energy Economics. IAEE annual conferences have been organised in Washington (1979),

Cambridge, UK (1980), at which the papers of this book were presented, and Toronto (1981).

BRITISH INSTITUTE OF ENERGY ECONOMICS

President: vacant
Vice-President: Mrs D.E.F. Carter
Chairman: Paul Tempest
Vice-Chairman: Eric Price
Treasurer: John Barber
Secretary: Niall Trimble
Committee: Robert Deam, Norman White, Philip Algar,
Anthony Scanlan, Gerry Corti, Walter Greaves, Christopher Johnson

Address: 9, St James's Square, London SW1Y 4LE.

Delegate List

*M.A. ADELMAN, MIT.
**J.A. ADAMSON, Chase Manhattan Bank.
*C. ANDERSON, EPRI.
**D. ANDERSON, Barclays Group of Banks.
E.J. ANDREW, B.P.
M. ANNESLEY, Tricentrol Oil Corp.
A.A. ATTIGA, OAPEC
B. BAHREE, Associated Press.
**J.M. BARBER, Dept. of Energy.
T. BARKER, University of Cambridge.
C. BARON.
**P. BARNES, Shell International Petroleum Co.
*P. BASILE.
R.W. BAUMGARTEN, IWG.
**P. BECK, Shell U.K.
**J. BEDORE, The Uranium Institute.
R. BEETHAM, College Retirement Equities Fund.

A. BEIJDORFF, Shell International Petroleum Co.
**R. BENDING, Energy Research Group, Cambridge.
I.D.G. BERWICK, UK Petroleum Industry Association Ltd.
**S. BILLER, European Parliament.
W. BLACKWELL, European Banking Co., Ltd.
**C. BRADLEY, Dept. of Industry.
L. BROOKES, UKAEA.
*B. BROWNE, Pacific Gas and Electric Co.
*I. BUPP, Harvard Business School.
C.P. CARTER, Chase Manhattan Bank.
**J. CARTER, Dept. of Energy.
A. CLIVE, Lazard Securities.
**I.B. COLLS, AERE.
N. COLLYNS, Scallop Corporation.
*M.A. CONANT, Conant & Associates Ltd.
R. CONDAP, Decision Focus Inc.

*Member of International Association of Energy Economists.
**Member of International Association of Energy Economists and member of the British Institute of Energy Economics.

**B. COOPER, Petroleum Economist.
D. COOPER, British Gas Corp.
P. COYNE, IPC Science & Techno-
logy Press.
**E. CROOK, British Petroleum.
R. DAFTER, The Financial Times.
C.P. DALTON, Petroleum Econo-
mics Ltd.
L. DARGIE, Fielding Newson Smith.
G.R. DARGIE, Bank of England.
*P.M.A. DAVIS, CPUC.
**R. DEAM (Queen Mary College,
London).
S.G. DIENER, Acres Consulting
Services.
A. DOWN, James Capel & Co.
*J. DUNKERLEY, Resources for
the Future.
*C.K. EBINGER, Project on Energy
and National Security.
C. EMBERSON, Energy Group,
Cambridge.
*P.L. ECKBO, The Chr. Michelsen
Institute.
*R.J. EDEN, Energy Research Group,
Cambridge.
K. ENDRES, Mobil.
R.T. EDWARDS, Bank of Scotland.
K. EGGE, Macalester College.
P.J. EHRLICH, Escola de Adminis-
traçã de Empresas de Sao
Paulo da Fundação Getülio
Vargas.
J.K. EVANS, Hampton Roads
Energy Co.
N. EVANS, Energy Research Group,
Cambridge.
T. EVANS, Hampton Roads Energy.
**N. FALLON.
I. FELLS, Dept. of Chemical Engin-
eering, University of Newcastle
J. FERGUSON, James Capel and Co.
A. FERRARI, C.E.A., Paris.
*A. FINIZZA, Atlantic Richfield Co.
D. FISHLOCK, The Financial Times.
*J. FOSTER, Petro-Canada.
R.E. FOX, Oil Exploration (Hold-
ings) Ltd.
*H.J. FRANK, University of Arizona.
J.T. FRASER, Tenneco.
M. FRITZ, Max-Planck-Institut.
T. GALLOWAY, Esso Petroleum Co
Ltd.
**C. GELVARIS, Motor Oil (Hellas).
**R. GIBSON-JARVIE.
**W. GREAVES.
J. GREENMAN, University of Essex.

*J. GREENWOOD, Charles River
Associates.
*M. GURFINKEL, Petroleos de
Venezuela.
**C. Hadjimatheou, S.S.R.C.
W. HAMILTON, Charter Consolidated
Ltd.
*J. HANSON, Exxon Corp.
S. HARRIS, Australian National
University, Canberra.
W. HAWTHORNE, Churchill College
Cambridge.
R. HEINE, Conoco Inc.
S.A. HESS, Security Pacific National
Bank.
L. HOFFMAN, Lehrstuhl für Volks-
wirtschaftslehre.
D. HOLTZ, General Motors Overseas.
C. HOPE, Energy Research Group,
Cambridge.
D. HOWELL, Secretary of State for
Energy.
**P. HUFFELMANN, Mobil Oil Co.
Ltd.
*J.P. HUGHES.
*W.R. HUGHES, Charles River Associ-
ates.
K.A.D. INGLIS, B.P.
**D. ION, World Petroleum Con-
gress.
J. IRVINE, Grenfell & Colegrave.
R. IRVING, Petroleos de Venezuela.
S. JHA, Administrative Staff College
of India.
A.R. JOHNSON, International Energy
Bank.
**C. JOHNSON, Lloyds Bank Ltd.
**C.L. JONES, British Embassy,
Washington.
**D. LE B. JONES, Dept. of Energy.
**T.P. JONES, Dept. of Energy.
*D.W. JORGENSON, Harvard Uni-
versity.
**M. KASER, St Antony's College,
Oxford.
M. KERR-DINEEN.
I. KING, Morgan Guaranty Trust
Co.
**P.O. KING, UKAEA.
L. KIRSCHEN, Associated Press.
P.H. KLINE, Dept. of Energy.
*R.F. KOSOBUD, University of
Illinois.
M. KOTAKE, Japan National Oil
Corp.
Y. KOVACH, Charter Consolidated
Ltd.

*A. LAJOUS-VARGAS, Secretaria de Patriuorio y Fourento Industrial.
M. LAPCO, Maraven SA.
D. LEDOUX, Royal Bank of Canada.
G.R. LEE, Shell International Petroleum Co.
J.R. LEMON
J. LEWINS, Cambridge University Engineering Dept.
*J.H. LICHTBLAU, Petroleum Industry Research Foundation Inc.
**F.C. LIPPINCOTT, Occidental International Oil Inc.
I.T. LOGIE, European Banking Co. Ltd.
**H. LOHNEISS, Morgan Guaranty Trust Company of New York.
R. LONG, IEA Coal Research.
A. LORIMER, British Nuclear Fuels Ltd.
M. LOVEGROVE, British National Oil Corp.
*F. LÜCKE, Federal Ministry of Economics, Bonn.
**R. MABRO, St Antony's College, Oxford.
**B. McBETH, Grenfell & Colegrave.
*A.S. MANNE, Stanford University.
*R. MASON, Petroleos de Venezuela.
J.F. MILLER, National Westminster Bank Ltd.
C. MILLS, Drayton Montagu Portfolio Management.
**J. MITCHELL, British Petroleum.
*M. NARKUS-KRAMER, The Mitre Corp.
*J. NAVARRO, Rand Corp.
R. NEESHAM, McGraw Hill.
B. NETSCHERT, National Economic Research Associates Inc.
D. NEWBERY, Churchill College, Cambridge.
*Ø. NORENG, Oslo Institute of Business Administration.
*A.N. NYSTAD, Chr. Michelsen Institute.
*P.R. ODELL, Erasmus University.
D. O'SHEA, M & G Investment Management Ltd.
**P. O'SULLIVAN, Welsh School of Architecture.
D. OSWALD, L. Messel & Company.
*R. PACHAURI, Administrative Staff College of India.

A. PACKER, U.S. Department of Labor.
M. PANIC, Bank of England.
**G. PARKIN, Walter J. Levy.
*A. PARRA, Petroleos de Venezuela.
F. PARRA.
**C.G. PEACOCK, Chem Systems International Ltd.
D.O. PEDERSEN, Danish Building Research Institution.
**J. PEGG, ICI Ltd., Petrochemicals Purchasing Group.
A.F. PEREIRA, Brazilian School of Public Administration.
P.R. PERKINS, Perkins Moody Associates.
M.K. PETERS, Brown Brothers Harriman & Co.
*J.L. PLUMMER, EPRI.
P.G. POPKIN, CREF.
**M.V. POSNER, Social Science Research Council.
**E. PRICE, Dept. of Energy.
G. PRICE.
W.K. PRYKE, Dept. of Energy.
J. PURVIS.
**H. QUICK, Shell International Petroleum Co.
P.C. QUINE, British Petroleum.
J. RAMPTON, Dept. of Energy.
*J. REDDINGTON, International Energy Agency.
**C. REID, European Investment Bank.
P. RODGERS, Sunday Times.
E.B. ROTH.
D. ROISSETTER.
**R. ROWBOTTOM, Midland Bank Ltd.
E.S. RUBIN, Energy Research Group, Cambridge.
M.L. RUDGE, Dupont Canada Inc.
E. RUTTLEY, World Energy Conference.
A. SAADIE, c/o Cities Services.
I. SAVILLE, Bank of England.
*J. SAVOY, Sun Co. Inc.
C.M. SAVY, Petroleos de Venezuela.
J. SAWHILL, Dept. of Energy, Washington.
**A.F. SCANLAN, British Petroleum.
S.W. SCHLICH, Bank of England.
D. SCHMITT, Universität Köln.
*G. SCHRAMM, The University of Michigan.
*S.H. SCHURR, EPRI.
G. SCOTT, Morgan Grenfell & Co.

J.G. SEAY, Institute of Gas Techno-
logy.
*D.B. SHEFFIELD, Stanford Univer-
sity.
**K.C. SHOVLAR, Shell International
Petroleum Co.
J. SILVERTOWN, Kings College,
Cambridge.
I. SKEET, Shell International Petrol-
eum Co.
D. SLADE, Bank of England.
T.J. SMALL III, Tesoro Coal Co.
**I. SMART, Ian Smart Ltd.
L.L.C. SMETS, Charter Consolidated
Ltd.
M.D. SMITH, Intercomp.
M. SPLINTER, Afd. Bedrijfskunde.
*D. SPRIGGS, Petroleum Analysis
Ltd.
J. STANISLAW, Energy Research
Group, Cambridge.
C. STARR, Electric Power Research
Institute.
D. STEEL, British Petroleum.
*W.A. STEGER, CONSAD Research
Corp.
*M. STEINBEISSER, Universite
Catholique de Louvain.
*D. STERNLIGHT, Atlantic Rich-
field Co.
P.H. SUDING, Universität Köln.
V.S.S. SURESH BABU, Bharat
Heavy Electricals Ltd.
J.L. SWEENEY, Stanford University.
C. TAYLOR, Department of Finance,
Ottawa.

*W.P.S. TAN, Nuclear Power Co.
Ltd.
*M. TELSON, House Budget Co.,
Washington.
**P. TEMPEST, Bank of England.
**F. THACKERAY.
J. THOMPSON, Fielding Newson-
Smith.
*L. THULIN, Den norske Creditbank.
T.P. TOWNSEND, Esso Europe Inc.
**N.E. TRIMBLE, British Gas Cor-
poration.
**L. TURNER, Royal Institute of
International Affairs.
*W.E. TYNER, Purdue University.
J. VALENTINI, Occidental Petrol-
eum Corp.
B. VAN VOORST, Time Magazine.
J. WADDAMS, Central Electricity
Generating Board.
**P. WARD, Cities Service S & T
Europe Africa Co.
*G.C. WATKINS, DataMetrics Ltd.
*L. WAVERMAN, University of
Toronto.
*J.B. WHARTON, Southern Califor-
nia Edison Co.
**N. WHITE, Norman White
Associates.
*J. WILKINSON, SKUN Co. Inc.
K. WILLIAMS, Shell International
Petroleum Co.
R. WILMOT, British Gas Corp.
*F. YOUNG, Conoco Inc.

Index